FRANKENSTEIN

AND THE

BIRTH

OF SCIENCE

JOEL LEVY

ANDRE
DEUTSCH

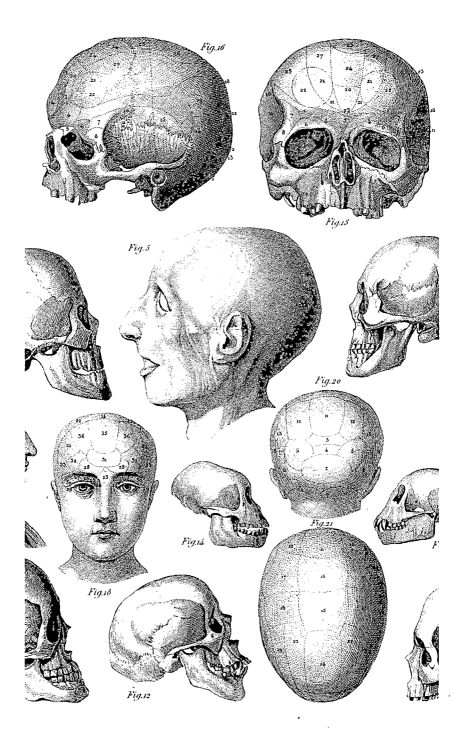

Fig.16

Fig.15

Fig.5

Fig.20

Fig.14

Fig.21

Fig.16

Fig.12

CONTENTS

INTRODUCTION

Mary Shelley's novel *Frankenstein: The Modern Prometheus* was published in 1818 with a preface authored anonymously by Percy Shelley, while a revised edition of 1831 included a much longer preface by Mary herself. Both prefaces give accounts – albeit passingly brief in the first – of the circumstances in which the novel was conceived. According to Mary's account, she and her lover (and soon-to-be husband) Percy Shelley were staying on the shores of Lake Geneva, near to Lord Byron and his physician John Polidori, when a period of dismal weather confined them to the house (*see* page 162). To amuse themselves, and having been reading German ghost stories, they agreed that they would each write a ghost story of their own. Mary recounts listening to conversations about contemporary advances in science regarding vitalism (*see* page 62) and galvanism (*see* page 40), and credits these with inspiring a waking reverie in which she saw "the pale student of unhallowed arts kneeling beside the thing he had put together ... the hideous phantasm of a man". Thus was born Frankenstein.

Literary and historical analysis throws into doubt almost every aspect of Mary's tale, but just as her novel itself has evolved into a myth, so her origin account has become a legend. Although it may be apocryphal, it encapsulates many of the themes of the age – themes that informed the novel and Mary's creations, and which thus feed into the extraordinarily rich confluence of sources that give the story its power and enduring resonance.

For this was the era of Romantic science, in which a new spirit of creativity and imagination, allied to a profound apprehension of the wonder of nature, blurred the boundary between natural philosophy and poetry, and saw a new breed of scientist emerge. Men like Humphry Davy and Samuel Coleridge (*see* pages 25 and 72) mixed science and poetry and drew from the same spring of inspiration for both; a synergy that would take fantastical flight in Mary's dark apotheosis of Romantic science.

The personalities and preoccupations of many of these figures would find expression in the fictional Victor Frankenstein, who rejects the dry objectivity of what he considers workaday science. Instead Victor embraces an approach that is physically and psychologically much more immersive and participatory – what today might be called a sort of "gonzo science" – and which is therefore fraught with danger and vulnerability.

In the person of the monster, or the creature – for Mary Shelley never gives Frankenstein's creation a name – are expressed an incredible range of the concerns, debates, doctrines and anxieties of the age. Mary's time followed on from the intellectual crises and revolutions of the Renaissance and the Enlightenment, which had overturned old authorities and certainties about man and his place in nature. Discoveries in biology, geography, geology and palaeontology, allied with the first stirrings of evolutionary theory, had lent impetus to a reassessment of man's relation to the natural world, and to the beast within.

But it would be the newest and oldest fields of natural philosophy that the novel would explore in the most thrilling fashion, and it is because of this that *Frankenstein* is often regarded as the first true work of science fiction. The novel and its protagonists embody and interrogate the promise and peril of fields such as electrical science and chemistry, psychology and the philosophy of consciousness. It is above all to these exciting and fascinating areas of science that the book speaks, and which this book in turn seeks to explain and explore.

A note on names:
Although not married when she conceived and began writing Frankenstein, *so that she was at this time more properly Mary Godwin, Mary married Percy in December 1816 and they considered themselves man and wife before this. The convention is thus to refer to her as Mary Shelley throughout the conception and writing of the novel. Unless specified or self-evidently otherwise, use of the name "Shelley" indicates Mary.*

CHAPTER 1

CHEMICAL
REVOLUTION

GETTING HIGH AND
LOSING YOUR HEAD

THE LATE EIGHTEENTH CENTURY WITNESSED AN
EXTRAORDINARY REVOLUTION IN THE WORLD OF CHEMISTRY,
AS IT TRANSFORMED FROM ALCHEMY INTO THE MOST
CELEBRATED SCIENCE OF THE AGE. THIS WAS A TIME WHEN
ROMANTIC ENTHUSIASTS WERE GETTING HIGH ON NEW
GASES, EVEN AS THE GREATEST SCIENTIST OF THE ERA BATTLED
FOR HIS LIFE WITH REVOLUTIONARY FANATICS – AND LOST.
MARY SHELLEY'S GREAT NOVEL *FRANKENSTEIN* EXPLORES THE
TRAUMATIC TRANSITION FROM THE OCCULT LORE OF ALCHEMY
TO THE PURE RATIONALISM OF CHEMISTRY.

DREAMS OF FORGOTTEN ALCHEMISTS: FROM ANCIENT MYSTICISM TO THE BIRTH OF A NEW SCIENCE

Mary Shelley wrote *Frankenstein* at a time of great excitement and optimism about science in general and about chemistry in particular. Chemistry represented the cutting edge of early nineteenth-century science: it was a youthful field of enquiry, taking its first steps among the more established sciences, such as physics and mathematics. But the novel aspects of chemistry imperfectly concealed its almost shameful former incarnation as the art of alchemy, with its inheritance of murky mysticism and shady occultism.

Shelley very deliberately frames her central character Dr Frankenstein at the intersection of these two worlds of chemistry; he is to be both a paragon of the new chemistry and an initiate of the old ways. His youth is spent "greedily imbibing" the works of the alchemists such as Paracelsus and Albertus Magnus: "fancies a thousand years old and as musty as they are ancient". Frankenstein reflects on how he "retrod the steps of knowledge along the paths of time and exchanged the discoveries of recent inquirers for the dreams of forgotten alchemists."

TO MOCK THE INVISIBLE WORLD

Arriving in the university town of Ingolstadt to advance his learning and be introduced to the world of natural philosophy, Frankenstein encounters two faces of the new chemistry – professors Krempe and Waldman. Krempe dismisses the alchemists as purveyors of "exploded systems and useless names", while scolding Frankenstein that "every instant that you have wasted on those books is utterly and entirely lost". Waldman is far more considered in his opinions, acknowledging the value of the groundwork laid by the alchemists, yet extolling the virtues of the brave new world of "modern chemistry": "The modern masters ... penetrate into the recesses of nature ... They have acquired new and almost unlimited powers; they can command the thunders of heaven, mimic the earthquake, and even mock the invisible world with its own shadows." Inspired by this vision, Frankenstein vows to master the modern science in order to achieve the ancient dreams of the alchemists – mastery over creation, and perhaps over death itself.

In Frankenstein, then, are embodied the tensions of the new chemical era, between its ancient baggage and its modern aspirations. The dreams and delusions of the alchemists, who sought the secret of the Philosopher's Stone and the Elixir of Life, set against the discovery of the new elements and airs, a new system of matter and its transformations.

Engraving from a book on alchemy, showing typically obscure symbolism; books of this kind inspired the fictional Victor Frankenstein.

THE DARK ARTS OF ALCHEMY

In her account of his introduction to university life, Shelley has Frankenstein receive a crash course in the evolution of chemistry – one that neatly recaps the history of the subject during its era of transformation, between around 1600 and 1800. The work and lives of those who played a part in this transformation offer tantalizing clues and teasing foreshadows of the story of *Frankenstein*.

Before chemistry there was alchemy, an occult art that sought to transform matter through manipulation of physical and mystical variables. The ultimate goal of the alchemist in the western tradition (China and other regions had their own, parallel and often similar traditions) was to create a magical substance or object known as the Philosopher's Stone, which had the power to transmute base metals into gold, among other wondrous attributes. In pursuit of this goal the alchemist combined ancient philosophy and metaphysics with more worldly attributes, such as the processing of metal ore and the preparation of spirits, which were derived in part from practical professions such as mining, tanning and dyeing. Thus, the alchemists of the medieval period named, isolated and purified a wide range of substances, laying essential groundwork for the science that would come later. As the great nineteenth-century chemist Justus von Liebig observed, "Without the Philosopher's Stone, chemistry would not be what it is today. In order to discover that no such thing existed, it was necessary to ransack and analyse every substance known on earth."

Among the alchemists specifically name-checked by Frankenstein is Paracelsus, the adopted moniker of the Swiss physician and philosopher Philippus Aureolus Theophrastus Bombastus von Hohenheim (1493–1541). A colourful character who claimed to have travelled the mystic Orient, Paracelsus made particular strides in pharmacy, introducing new treatments for syphilis and developing laudanum, a tincture of opium that proved to be the most beneficial tool in the physician's pharmacopoeia right up until the mid-nineteenth century. More generally, Paracelsus' brave assault on the prevailing deference to received authority did much to open the door to a new spirit of experimental philosophy, inspiring the philosophers who would later bring about the Scientific Revolution.

PARACELSUS AND THE HOMUNCULUS

Of particular interest in the light of *Frankenstein* the novel, Paracelsus is famous for his claim to have successfully created a homunculus – literally, "little man" – an artificial being, precursor of the Jewish *golem* and Frankenstein's monster. According to the recipe Paracelsus came up with, if sperm, "enclosed in a hermetically sealed glass, is buried in horse manure for forty days, and properly magnetized, it begins to live and move. After such a time it bears the form and resemblance of a human being, but it will be transparent and without a body." Paracelsus further claimed that feeding it an extract of blood named *arcanum sanguinis hominis* would result in a tiny human child that could be raised in the normal way.

With his outlandish claims and mystical doctrines, Paracelsus still belonged to the world of the alchemists. The figure that truly bridged the divide between ancient mysticism and modern science was the Anglo-Irish natural philosopher, Robert Boyle (1627–1691). Boyle was initiated into the lore of alchemy, but strove to redirect the ancient art into a new, more scientific direction. However, Boyle himself was not above fantastical whimsies and speculations – some of which appear to anticipate elements of the *Frankenstein* story. In a kind of mission statement for future scientists that he recorded for the Royal Society, the learned body he helped found in London to further the cause of experimental science, Boyle listed a number of avenues for future research. Alongside encouragement to develop mind-altering drugs and pain relief, Boyle also recommended researches aimed at "Attaining Gigantik Dimensions" – which was believed to be a reference to the possibility of enlarging the human race.

THE BREATH OF LIFE: PNEUMATIC CHEMISTRY AND NEW AIRS

Part of the Romantic appeal of chemistry was its exciting and often daring research into what were then known as "airs" – or gases. This made a huge impact on the scientific imagination of Mary Shelley and hence fed into the story of *Frankenstein*. The field sometimes known to Shelley's contemporaries as "pneumatic chemistry" (from the Greek word *pneuma*, meaning "breath") isolated and identified a hitherto unheralded phase of matter – and in doing so, explored the breath of life itself. Scientists experimented with newly discovered gases that could end life, enhance life and even prolong life, opening the doors of perception and promising to reveal hidden mysteries of vital phenomena.

VON GUERICKE'S SPHERE

A dramatic illustration that mere airs could exert tremendous force came with the celebrated Magdeburg Sphere experiments. In 1650, Otto von Guericke (1602–86), a German military engineer and the mayor of Magdeburg, constructed one of the first air pumps. In 1654, in a demonstration conducted in the presence of the emperor Ferdinand III, von Guericke fitted together two giant copper hemispheres and pumped all the air out of the enclosed space. Despite the fact that no screws or ties held the domes together, they could not be pulled apart by a team of 16 horses. Only when von Guericke released a valve to re-admit the air did the hemispheres fall apart. This was science as theatre, a theme that would become increasingly important in the Romantic era of science, exerting a force of its own on the imagination of both authors and audience.

AIR POWER

The notion that "airs" – the vague label applied by alchemists to fumes, vapours and the atmosphere itself – might play an important role in chemistry, and particularly the chemistry of life, was suggested as early as the fifteenth century by the work of Nicholas of Cusa (1401–64). Nicholas was a German theologian and natural philosopher who carefully weighed a potted plant at intervals, proving that it gained weight with no input but the air. This was considered to be the first modern formal experiment in biology and was among the first proofs that air itself had weight.

THE POWDER OF SYMPATHY

Following in the footsteps of Nicholas of Cusa was an alchemist who would probably have been on Frankenstein's reading list (*see* page 12): Jan Baptista van Helmont (1579–1644). Van Helmont is known today as a pioneer of pneumatic chemistry, but in his own time he was better known for a controversy over a strange magical medicine known as the "Powder of Sympathy".

A Flemish physician and alchemist from a noble family, van Helmont studied at Louvain and travelled extensively around Europe. However, he preferred to retreat to his country estate in order to pursue his researches, which were both scientific and concerned with the occult. (In fact, such a distinction hardly existed at this time: both were simply branches of natural philosophy.) Despite his retiring nature, van Helmont managed to attract religious censure for his belief in the efficacy of a strange ointment that was made from the skull of a man who had died a violent death, boar and bear fat from animals killed while mating, burnt worms, dried boar's brain, red sandalwood and powdered mummy. This concoction was known as the Powder of Sympathy, for it was claimed to operate on the magical principle of "sympathy" – the supposition that objects or substances, once associated, retain influence over one another.

Following Paracelsus (*see* page 14), van Helmont believed that the powder could be used to treat wounds – by applying the ointment to the blade that had inflicted them! The patient need not even be aware of the application, and the magical cure could supposedly be applied from a great distance. For his support of this controversial treatment, van Helmont was sentenced to house arrest.

Flemish alchemist and physician Jean Baptiste van Helmont, one of the fathers of pneumatic chemistry and a biochemistry pioneer.

THE WOMB OF THE WATERS AND THE SPIRIT OF THE AIR

Van Helmont is most celebrated for two experiments in which some of the earliest identification of airs went hand in hand with the first explorations of what today would be called biochemistry. The first was a carefully controlled and fastidiously measured version of the pot-plant experiment of Nicholas of Cusa, in which van Helmont weighed a pot and the plant it contained before and after growing it for five years, during which it was fed only distilled water. Finding that it had gained 74kg (164lb) in that time, van Helmont concluded that: "164 pounds of Wood, Barks, and Roots, arose out of water only".

Van Helmont believed that water was the chief constituent of matter, and thus that the plant had gained mass by converting water into biomass. "The whole rank of Minerals," he wrote, "do find their Seeds in the Matrix or Womb of the Waters ..." His wording reflected contemporary views of mineral, vegetable and animal as related phenomena in which forms of vital action could operate without distinction. In other words, the line between animate and inanimate was blurred in a way that would continue to inform perceptions in Mary Shelley's lifetime.

This belief was reflected in the terminology that van Helmont applied to the vapour or airy spirit he showed had been driven off from charcoal during combustion. With a derivation based on the Greek word *chaos*, van Helmont wrote: "I call this Spirit, unknown hitherto, by the new name of Gas, which can neither be constrained by Vessels, nor reduced into a visible body". To the gas produced from burning charcoal, he gave the name *spiritus sylvester* ("spirit of the wood") – what we now call carbon dioxide.

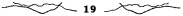

Frontispiece illustration of a book on the Powder of Sympathy, the mystical ointment researched by van Helmont and other early proto-scientists.

BREATH MADE VISIBLE

Through the exploration of gases, scientists approached ever closer to investigation of the biochemistry of life. For instance, a subsequent landmark in pneumatic chemistry came in the form of a dramatic demonstration performed by the Scottish scientist Joseph Black (1728–99), in which he proved that he had isolated and identified a gas contained in the very air that we exhale.

In a series of experiments in the years 1754–6, through heating chalk (calcium carbonate) to produce quicklime (calcium oxide), Black obtained what he called fixed air (since the gas was contained within the solid until released by heating). It was the same gas described by van Helmont a century before, but Black was able to show that it was the product not just of combustion, but of respiration – the process of breathing and one of the key phenomena of life. He first prepared a solution of slaked lime (aqueous calcium hydroxide), known as limewater, and then exhaled into a jar of it. The carbon dioxide in his breath reacted with the dissolved calcium hydroxide to form particles of chalk (calcium carbonate), so that the clear solution turned a milky white.

Joseph Black did pioneering work on pneumatic chemistry, and also on the physics of heat and energy transference.

The son of a wine merchant, who carried out many of his researches in a brewery, Black explored the products and processes of fermentation, and was able to show that the same "air" linked diverse natural phenomena: breathing, fermentation and combustion. Pneumatic chemistry was lifting the veil of mystery surrounding the vital processes.

Black's experiments extended as far as exploring the effects of his new "fixed air" on the boundary between life and death, foreshadowing Joseph Priestley's experiments with mice (*see* page 24). "In the same year," he later wrote, "in which my first account of these experiments was published, namely 1757, I had discovered that this particular kind of air ... is deadly to all animals that breathe it by the mouth and nostrils together; but that if the nostrils were kept shut, I was led to think that it might be breathed with safety. I found, for example, that when sparrows died in it in ten or eleven seconds, they would live in it for three or four minutes when the nostrils were shut by melted suet." Not for the last time the public could read about a scientist playing God.

POP PIONEER

Among the first to isolate and describe the gas that is known today as oxygen was Joseph Priestley (1733–1804), a radical non-conformist with a talent for brilliant experiments. Several aspects of Priestley's accomplished career tantalizingly foreshadow that of the fictional Frankenstein.

A minister and teacher, Priestley pursued scientific researches after being encouraged to do so by Benjamin Franklin (the American electrical pioneer and one of the Founding Fathers of the United States – *see* page 37). Like Black he did some of his most important work in a brewery; it was in the one next to his house, for example, that he proved the gas bubbling up to the surface of the vats of fermenting beer to be the same as Joseph Black's "fixed air" (that is, carbon dioxide). The presence of a ready supply of the gas encouraged him to simulate the natural effervescence of some mineral waters, which he did by dissolving the carbon dioxide under pressure in water, thus creating carbonated water and setting off a European craze for "soda water".

*Equipment including the pneumatic
trough used by Priestley to capture,
isolate and experiment on new airs or gases.*

AIR REPAIR BY VEGETABLE CREATION

Priestley moved on from carbon dioxide to perform a series of elegant and fascinating experiments with a new, as yet unnamed air. This appeared to be intimately related to the phenomena of life, such as the growth of plants and the breathing of animals. In August 1771, Priestley made one of the earliest discoveries relating to the biochemistry of photosynthesis. Enclosing a burning candle inside a bell jar (a sealed glass vessel) alongside a mint plant, Priestley observed that the candle soon guttered and went out. After waiting 27 days, he used his "burning glass" – an arrangement of two lenses that could concentrate the rays of the sun to produce a point of intense heat – to relight the candle wick, whereupon it burned happily. No new air had been admitted into the bell jar, so Priestley was able to conclude that the green plant had somehow replenished the combustible capacity – or ingredient – of the air. It was known that combustion and respiration (the physiological activity of animals) involved similar chemistry and gave the same product (Black's "fixed air", or carbon dioxide), and Priestley realized that he had uncovered some vital principle of life with global consequences, surmising that "the injury which is continually done [to the ability of the air to support respiration] by such a large number of animals is, in part at least, repaired by the vegetable creation."

PRIESTLEY'S MOUSE

Three years later, Priestley's most celebrated experiment would go further in exploring the link between the breath of life and this combustible air (which would later be named oxygen), produced by plants. Using his burning glass, Priestley heated an oxide of mercury and collected the gas that was driven off; he found it to be colourless and odourless, causing a flame to burn very brightly. Further testing revealed that it was "superior" to common air, for as he related:

> I procured a mouse, and put it into a glass vessel, containing two ounce-measures of the air ... Had it been common air, a full-grown mouse, as this was, would have lived in it about a quarter of an hour. In this air, however, my mouse lived a full hour ... and appeared not to have received any harm from the experiment.

This passage reveals a scientist using cutting-edge discoveries seemingly to defy death – or at least postpone it – and playing God with a living organism. The similarities between Priestley and Frankenstein do not end there, for the unfortunate Priestley would eventually suffer a similar fate to countless screen iterations of Frankenstein: he was driven from his home by a torch-wielding mob, and forced to flee while they burned his laboratory to the ground.

A torch-wielding mob destroys the Birmingham home of chemist Joseph Priestley, foreshadowing the fate of many cinematic versions of Victor Frankenstein.

THE AIR IN HEAVEN: CHEMISTRY INTOXICATES THE ROMANTICS

The next Briton to pick up the torch of pneumatic chemistry was Humphry Davy, the most famous scientist of his age and a figure who looms large in the cultural and scientific hinterland of *Frankenstein*. As discussed in Chapter 7, Davy must be considered one of the primary inspirations for the fictional personage of Frankenstein himself, despite the apparent disparity between the dapper little provincial gent of reality and the brooding, Byronic figure of the book. Davy came from a lower middle-class background in Cornwall, in the far southwest corner of England, but with intense energy and ambition had taught himself chemistry through books and an apprenticeship with an apothecary. Among his many claims to fame were his electrochemical researches (*see* page 49) and invention of the miners' safety lamp, but it was in an earlier phase of his career, during his time at the Pneumatic Institute in Bristol, that he arguably set the template for the kind of immersive gonzo science that would lead Frankenstein astray.

A poet as well as a scientist, Humphry Davy has often been seen as one of the models for Victor Frankenstein, although in practice it is Professor Waldman who is based on him.

DANCING AND VOCIFERATING

The Pneumatic Institution for Relieving Diseases by Medical Airs was a philanthropic venture in Bristol, in the west of England, set up by an eccentric physician named Dr Thomas Beddoes in 1798. He believed that inhalation dramatically increased the medicinal potency of gases, but he was a little hazy on which of the increasingly long menu of recently discovered airs had therapeutic value.

To explore this issue he recruited Davy as Medical Superintendent, and in 1799 the young tyro set about testing various gases, primarily on himself. His adventures in the world of bioactive gases – and specifically psychoactive ones – became touchstones for the new Romantic science, blending experimental science with psychonautic exploration. It was brave, potentially foolhardy and dangerous work, as when Davy found himself "sinking into annihilation" after inhaling carbon monoxide.

Davy identified nitrous oxide as one of the most promising of the "medical airs" and proceeded to experiment with ever-greater doses, while monitoring his psychological and physiological response. Today, nitrous oxide is often known as laughing gas, and Davy soon experienced its euphoric effects: "Sometimes I manifested my pleasure by stamping or laughing only; at other times, by dancing around the room and vociferating." In May 1799, Davy stumbled upon a new realm of anaesthesia, noting that inhaling over six litres of nitrous oxide produced an "experience [that] for a moment so intense and pure as to absorb existence. At this moment, and

DAVY AND THE LAWS OF LIFE

Some of Davy's earliest writings set out materialist principles that foreshadowed *Frankenstein*'s guiding conceit that inanimate matter could be endowed with mental powers and even a soul. In an essay of 1798 he had insisted that "the laws of the mind ... are not different from the laws of corpuscular motion [that is, physics]", a for-the-time radical materialist position offering the promise that "[by] experimental investigation [into] the organic matter of the body ... we should be informed of the laws of our existence ... Thus would chemistry, in its connection with the Laws of Life, become the most sublime and important of all the sciences."

not before, I lost consciousness; it was, however, quickly restored ..." Davy later wrote about the potential of the gas for medical anaesthesia, but never pursued this line of enquiry, which was not taken up for nearly half a century.

A satirical depiction of the craze for laughing gas: Georgian scientific demonstrations straddled the border between education and disreputable entertainment.

THIS WONDER-WORKING GAS

Instead, Davy plunged deeper into the psychoactive possibilities of nitrous oxide, constructing a sort of gas chamber to enable inhalation of massive doses. In one notorious session in December, he became "completely intoxicated" after inhaling 57 litres, experiencing an intense hallucinogenic trip, after which he "stalked majestically out of the laboratory" to inform Dr Robert Kinglake "with the most intense belief and prophetic manner ... 'Nothing exists but thoughts!—the universe is composed of impressions, ideas, pleasures and pains!'"

The gas thus offered value both for recreation and something more profound: a key to the doors of perception, opening a new world of creative inspiration and psychic exploration. Davy introduced it to his circle of poetic friends, among whom it created a tremendous stir. The poet Robert Southey extolled, "I am sure the air in heaven must be this wonder-working gas of delight." Writing to his brother he reported breathlessly, "Davy has actually invented a new pleasure, for which language has no name. I am going for more this evening!"

Though encountered at some years' remove, these accounts of intoxication must themselves have been intoxicating to the young Shelleys, both of whom encountered Davy in print and in person. He was, for instance, a friend of Mary's father, and as a 14-year-old she attended lectures given by Davy at the Royal Institution in London, while in the novel she would borrow almost verbatim one of his panegyrics about the prospects of the new chemistry and the men of science exploring it, putting it in the mouth of the professor, Waldman. Furthermore, Mary's journal of 1816 reveals that even as she was in the thick of writing *Frankenstein*, her reading matter at the time was Davy's *Elements of Chemical Philosophy*.

HYDROGEN AND BALLOMANIA

One of the most emblematic manifestations of Romantic chemistry was its role in ballomania – the craze for ballooning that swept Europe in the late eighteenth century. The isolation of hydrogen, and the discovery of its extreme buoyancy, had inspired speculations about lighter-than-air devices, speculations that bore spectacular fruit in 1783, when Dr Jacques Alexandre Charles and an assistant ascended into the skies above Paris in a hydrogen balloon, attracting a crowd of 400,000 (half the population of Paris). Benjamin Franklin, the American ambassador in Paris, memorably reported that, "Someone asked me – what's the use of a balloon? I replied – what's the use of a newborn baby?" Ballooning showed how Romantic science could inspire new frontiers in human endeavour, combining exploration of uncharted regions with cutting-edge technology – a similar recipe to that achieved by Mary Shelley, and characteristic of the spirit of the age that inspired her. It is notable that Mary constructed a small hot air balloon for Percy as a birthday treat.

Although the Montgolfier brothers had beaten him to manned flight, it was Professor Charles who would make the most prodigious and revelatory flight, using a hydrogen balloon.

CHAPTER 2

ELECTRIC FLUIDS AND ANIMAL SPIRITS

GALVANISM, VOLTAIC PILES, ELECTROCHEMISTRY AND THE START OF A NEW ERA

THE SHARPEST EDGE OF LATE EIGHTEENTH-CENTURY SCIENCE WAS THE MAGICAL, ALMOST SUPERNATURAL WORLD OF ELECTRICITY. IT WAS A TIME OF STUNNING DEMONSTRATIONS – SUCH AS ATTEMPTS TO REANIMATE CORPSES FROM THE SCAFFOLD – AND OF DIY EXPERIMENTS THAT CAUGHT THE IMAGINATIONS OF YOUNG DREAMERS LIKE MARY AND PERCY SHELLEY, STIRRING THEM TO FEBRILE SPECULATIONS ON THE POSSIBILITIES OF THIS NOVEL FORCE OF NATURE.

DANGER! HIGH VOLTAGE! ELECTRICAL FIRE AND ELECTROSTATIC GENERATORS

The story of electrical science – the field that more than any other must have stirred the imagination of Mary Shelley as she conceived her monster – begins in antiquity, in much the same place that the young novelist would later encounter it: with bioelectricity (electricity in living organisms).

FISHERMAN'S FIEND

Among the earliest written records of electricity of any kind is an ancient Egyptian hieroglyphic mention of a creature called a "sheatfish", describing how it "releases the troops". This is believed to be a reference to an electric catfish capable of generating shocks of more than 450 volts, thus forcing ancient fishermen to release their catch or risk electrocution from a single sheatfish caught in the net.

The word "electricity" itself is derived from the ancient Greek word for amber (fossilized pine resin), which was known to acquire strange properties if rubbed with a cloth or leather. Amber prepared in this way could pick up small objects, and, viewed in darkness, would produce flashes of light. Ancient and medieval writers including Pliny the Elder and Giovanni Battista della Porta described such phenomena, but no systematic enquiry was made until the English physician and natural philosopher William Gilbert (1544–1603) published his remarkable book *De Magnete* (On Magnets), in 1600. Gilbert coined the term "electrica", from the ancient antecedents, to describe the attractive force generated by rubbing certain substances. Following on from this, the English natural philosopher Thomas Browne would use the word "electricity".

THE SULPHUR BALL

Gilbert's key experimental tool was his *terrella*, a globe of lodestone (magnetic ore). Possibly inspired by this, the German inventor Otto von Guericke – of vacuum fame (*see* page 16) – invented his "sulphur ball", the first electrostatic generator, in around 1660. This was a large sphere of sulphur that sat in a wooden cradle and could be rotated around a central rod. When rubbed by hand as it spun, it would acquire a static electric charge, which could be used for experiments.

The sulphur ball itself was created by pouring molten sulphur into a hollow glass sphere, which was smashed after the sulphur had cooled. However, at some point it was discovered that the glass sphere on its own was at least as effective at holding charge.

Von Guericke and his sulphur ball electric machine, one of the first electrostatic generators.

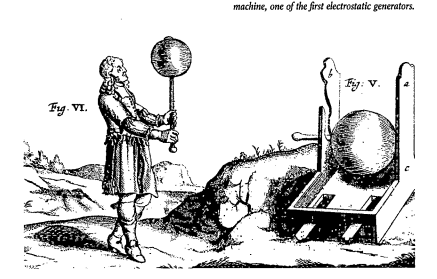

A further advance came after Evangelista Torricelli invented the mercury barometric device – a long tube, open at one end, which was filled with mercury and then inverted in a pool of mercury, creating a vacuum gap at the top of the tube. When the tube was shaken and viewed in the dark, a glow of light could be observed at the vacuum end.

In 1709, this discovery inspired the English natural philosopher Francis Hauksbee to create a rotating glass globe from which all the air had been evacuated. When he pressed his hand against it as it spun round, it lit up with a glow bright enough to read by. This was possibly the first time in history that someone could read by a non-combustible artificial light source, and Hauksbee's device is the direct antecedent of modern plasma globes. It is also perhaps the first in a line of striking and dramatic electrical technologies of the sort that would later come to be associated with Frankenstein-like creation scenes.

The "Electric Kiss", a popular and somewhat risqué application of electrostatic technology; the device is cranked to generate a charge, which passes through the lady via her lips.

THE ELECTRIC KISS

As electrostatic generators became increasingly available, electricity was widely known as an engaging curio. A popular salon pastime, for instance, was the "Electric Kiss" – a demonstration of how electrostatic charge could pass from one individual to another on contact. Increasingly large charges could be generated, but could they be stored?

In 1745, the German clergyman and scientist Ewald Jürgen von Kleist decided that the natural place to store electricity – which was widely conceived of as a fluid – would be in a bottle. He discovered that a jar or bottle filled with water or mercury, sat on a metal base, could indeed store charge: enough to knock him down! An almost identical device was invented at the same time by the Dutch physicist Pieter van Musschenbroek, at the University of Leiden, and so came to be known as the Leiden or Leyden jar. Demonstrating the power of the device, van Musschenbroek administered such a tremendous shock to a student named Andreas Cunaeus that the unfortunate man insisted that he would not repeat the experience for the whole kingdom of France. Leyden jars can indeed be dangerous; a jar with a capacity of just 1/2 litre (1/8 gallon) can deliver a fatal shock.

FRANKLIN'S KITE

The increasingly large and powerful sparks that could be drawn from such devices prompted consideration of parallels with lightning, and in one of history's most famous and mythologized experiments American polymath Benjamin Franklin (1706–90) attempted to prove that they were one and the same. Franklin had already speculated that lightning is a form of "electrical fire", and proposed attracting it using a long metal rod extending from the ground to the top of a tall building. At the time he lived in Philadelphia, where no such building then existed, so he claimed in a letter to have hit upon a short-cut, detailing an experiment in which he flew into a storm cloud a kite to which a metal key was attached. The key, he claimed, attracted a lightning strike and stored the resulting charge, which could be downloaded into a Leyden jar and even felt as a tingle. Since this experiment is incredibly dangerous, there has been widespread scepticism over whether Franklin genuinely performed it. One theory is that he invented it as a kind of revenge prank on a British scientist accused of stealing his ideas.

By the time he supposedly flew his kite into the storm, Franklin's original proposal for a wire lightning conductor – which had been outlined in a 1750 letter to Europe – had prompted European scientists into successful trials. On 10 May 1752 the French naturalist Thomas-

François Dalibard used a large metal rod to conduct electricity from lightning, thus proving Franklin's contention that lightning is a form of electricity.

Franklin wrote extensively on electricity, coining several important terms including battery, charge, conductor, plus, minus, positively, negatively and condenser (another term for a capacitor, such as a Leyden jar). But he would remain best known for his achievement of "bringing lightning from the heavens", as his famous experiment was described by British scientist Joseph Priestley in a widely read account of 1767. Priestley was a friend of Mary Shelley's father, William Godwin, and it is likely that she would have read of this famous experiment. In the novel there is no specific mention of electricity in the creation of the monster, let alone the scene so familiar from movie versions in which a lightning strike electrifies it into life, yet there was nonetheless a potential transmission of ideas from Franklin's kite to Mary's imagination.

FRANKLIN AND THE YEAR WITHOUT A SUMMER

Benjamin Franklin fitted a lightning rod to his house and used it for experimental observations; he even fitted it with a series of bells that would sound when a strike was successfully drawn. Other properties in the city were also fitted with lightning rods. In 1816, the year without a summer that helped engender the creation of *Frankenstein*, the citizens of Philadelphia developed a bizarre conspiracy theory closely resembling modern anxieties about the atmospheric experiments of the High Frequency Active Auroral Research Program (HAARP) in Alaska. Seeking an explanation for the unseasonable weather, which included snow falling in July, they blamed Franklin's lightning rods for affecting the weather and somehow reversing the seasons.

SHOCK TO THE HEAD

Franklin's dangerous experiment to capture lightning in a bottle was performed successfully on at least one occasion, but it was also responsible for the first recorded victim of high-voltage experimentation. In 1783, the Baltic German physicist Georg Wilhelm Richmann (or Reichmann) was killed while trying to charge Leyden jars from a lightning strike in St Petersburg. Perhaps the earliest attested example of ball lightning travelled down the wire from the lightning conductor and struck Richmann in the head, burning his lungs and one of his shoes and blowing the laboratory door off its hinges.

An engraving of the fatal encounter between ball lightning and Dr Richmann, during his ill-fated attempt to recreate Franklin's kite experiment.

A SPARK OF BEING: GALVANISM AND THE SECRET OF LIFE

What power animates the limbs? What energy drives the heart to beat? What medium conducts the will from the mind to the hands or the feet? Is there some vital force or spirit at work in the body, and can it be isolated and interrogated ... perhaps even reproduced? Questions such as these intrigued natural philosophers from ancient times to the Enlightenment, with the late eighteenth century in particular bringing new discoveries to overturn old certainties, as advances in electrical science hinted at an awesome possibility: the potential for humankind to create and control the vital force that animated life.

NEUROMUSCULAR ACTION

The mechanism of muscle activation and control – what today we would call neuromuscular action – was attributed by Galen, the leading Classical authority on physiology, to a kind of fluid that he called "animal spirits". Galen's influence endured throughout the Middle Ages, and in the seventeenth century his theory was brought up to date by the French philosopher René Descartes (1596–1650). He proposed a hydraulic theory of neuromuscular action, in which the role of the nerves is to conduct fluids between the brain and the muscles. In his 1648 *Traité de L'Homme* ("Treatise on Man") he wrote:

> ... one can well compare the nerves of the machine that I am describing to the tubes of the mechanisms of these fountains, its muscles and tendons to diverse other engines and springs which serve to move these mechanisms, its animal spirits to the water which drives them, of which the heart is the source and the brain's cavities the water main.

Some contemporary anatomists agreed with Descartes; the seventeenth-century English anatomist Thomas Willis, for example, believed the role of the brain to be the "conversion of vital spirits ... into essential animal spirits". But a 1664 experiment by Dutch scientist Jan Swammerdam destroyed one of Descartes' central contentions. By sealing a frog muscle in a glass vessel from which projected a narrow tube holding a water droplet, Swammerdam was able to test whether the overall volume of the muscle changed during contraction. When he stimulated the muscle with a silver wire attached to a copper loop, the water drop did not move: the muscle volume had not changed, in direct contradiction of Descartes' hydraulic theory.

What was particularly interesting about Swammerdam's experiment – although he did not pursue it – was his use of a bi-metallic instrument to stimulate the muscle. This may have been the first instance of external electrical stimulation of neuromuscular action, but it is also possible that Swammerdam mechanically stimulated the muscle. It was a foretaste of the landmark research to come.

Even as Descartes' hydraulic theory wavered, an alternative came into focus. The invention of static electricity generators meant that an electrical charge could be produced at will, while the development of the Leyden jar, a device capable of storing charge and discharging it on demand (*see* page 37), offered researchers a powerful new tool for exploring the remarkable potential of electricity.

SHOCKING DISCOVERIES

Instances of bioelectricity provided an obvious link between electricity and the mysterious energies that triggered and powered neuromuscular action; the shock dispensed by a Leyden jar was compared to that of the electric eel or torpedo (electric ray). Electricity was widely conceptualized as some sort of ethereal fluid, which chimed with the belief that neuromuscular action was mediated by some sort of "animal spirits" or "nervous fluid". Albrecht von Haller, professor of anatomy and medicine at Göttingen University in Germany, speculated that the nature of the "nervous fluid" might be due to "electrical matter" present in the "animal spirits".

To investigate this potential link, Luigi Galvani (1737–98), professor of anatomy at Bologna University in Italy, used frogs' legs as a testing platform in a programme of research starting around 1781. He dissected away the upper halves of frogs to leave pairs of legs with a portion of the spinal cord projecting above them, and looked at how they reacted when they were hooked up to electrostatic generators and Leyden jars by various arrangements of wires attached to the nerves. However, it was only when his assistant (probably his wife, Lucia) touched a nerve with a scalpel just as an electrical spark was discharged nearby, that Galvani achieved a notable result: "All the muscles of the leg seemed to contract again and again as if they were affected by powerful cramps." Galvani speculated that the "electrical atmosphere" (what today would be called the electrical field) caused by the spark had stimulated movement of the electrical fluid contained in the frog's nerves.

LO AND BEHOLD

To investigate further, Galvani experimented with frogs' legs hooked to the iron railings of his garden. Following the experiments of Benjamin Franklin, which had demonstrated lightning to be a form of electricity, Galvani wanted to see if nearby lightning strikes or other atmospheric electrical effects would affect his frogs' legs. He achieved some positive results, but was more excited by the finding that the legs sometimes twitched in the absence of storms or even clouds: "lo and behold, the frogs began to display spontaneous, irregular and frequent movements".

To Galvani, this suggested that the muscles themselves were the source of the animating electrical fluid. In particular he showed that contractions could be elicited by touching the legs with a bi-metallic arch, in exactly the same fashion that a Leyden jar is discharged. For Galvani this was strong evidence that a muscle and its nerves act like a biological Leyden jar: the nerves conducted some sort of electrical fluid to the outer surface of the muscle, equivalent to a conductor carrying charge to the outer surface of a Leyden jar. This caused the inner surface to become oppositely charged, with the opposition leading to muscular contraction.

THE DEMONSTRATED TRUTH OF ANIMAL ELECTRICITY?

Galvani soon fell into a protracted dispute with Alessandro Volta (1745–1827), a professor at the neighbouring University of Pavia and the leading authority on electrical science. Volta, who described himself as having "a genius for electricity", had built his first lightning rod at age 17 and later invented the "perpetual electrophosphorus" (a static electricity generator), securing himself the chair of experimental physics at Pavia at the age of just 33.

Despite initial doubts about Galvani's findings, he repeated the experiments for himself and changed his position "from incredulity to fanaticism", as he put it himself. The discovery "proves animal electricity", he enthused, "and places it among the demonstrated truths". However, further tests showed that the muscles could be induced to contract purely by touching the bi-metallic contacts to the nerves, without touching the muscles at all, and this appeared to undermine Galvani's Leyden jar analogy. Volta began to suspect as more appropriate a quite different analogy: between the neuromuscular unit and an electroscope – a primitive device for detecting electrical current. Volta proposed the frog muscle to be detecting an external electrical potential, rather than electricity resident, as Galvani claimed, in the biological tissue itself.

Where was the electric current coming from? Volta had found that using a probe composed of just one metal produced no result; a bi-metallic probe seemed to be essential. He proposed that contact potential – a phenomenon whereby electrical current is generated by contact between two metals, discovered by German mathematician Johann Georg Sulzer in 1752 (see page 46) – was responsible for the electricity.

Volta's critique of Galvani's results saw the scientific community divided into "animalists" – those who supported Galvani's claim that the electricity driving the muscle contractions is biological in origin – and "metalists" – those who supported Volta. It was in refuting the claims of the animalists and seeking to create an entirely non-biological model of the current generation that Volta was led to create the battery, an invention that would prove to be a key element of Frankenstein's technological inheritance (see page 45).

In fact, both metalists and animalists were right. Volta had correctly identified contact potential in the bi-metallic arch as the source of electricity in Galvani's experiment, but Galvani's beliefs about animal electricity would later be proven right. Muscular contractions are indeed stimulated by electricity transmitted through the nerves, and living tissue can generate electricity.

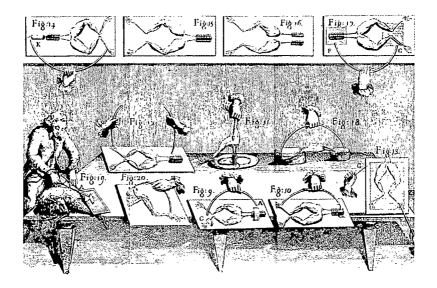

Galvani at work in his rather ghoulish laboratory. But was the source of the electrical energy the metallic arch or the animal tissue itself?

THE INFLUENCE OF GALVANISM

Despite the dispute and doubts over Galvani's hypotheses, his work became widely known and celebrated. Science had demonstrated a direct link between electricity and the energies vital to life, and even, apparently, to have reanimated dead tissue. Perhaps inevitably, attempts were made to apply Galvanism to human corpses, with some remarkable and creepy results (*see* page 53). Galvani's work and laboratory also made for a striking, archetypal image: surrounded by dismembered corpses, grisly arrangements of muscles and nerves and strange instruments emitting sparks and crackles, the scientist dared to probe the mysteries of life and death.

Certainly Mary Wollstonecraft Shelley was familiar with Galvani's discoveries and their potential applications: she later recalled discussing with Byron and Percy Shelley that: "Perhaps a corpse would be re-animated; Galvanism had given token of such things." The key passage in *Frankenstein*, in which the protagonist animates his creation, has clear echoes of Galvani's experiments: "... I collected the instruments of life around me, that I might infuse a spark of being into the lifeless thing that lay at my feet ... a convulsive motion agitated its limbs." Just as Galvani's work must have helped inspire Frankenstein's own efforts, so it seems plausible that the man himself may have helped inspire Mary's protagonist. Certainly, nearly all subsequent attempts to realize Frankenstein's dream in real life – and make a human or at least a simulacrum of one – have relied on Galvani's breakthrough discovery and the phenomenon of neuromuscular electricity.

THE KEY TO UNLOCK NATURE: ZOMBIE TONGUES, VOLTA'S PILE AND THE ELECTROLYSING EMPEROR

E ven with the invention of the electrostatic generator and the Leyden jar (*see* page 37), electricity had remained something of a wild card – unpredictable and uncontrollable. Only with the invention of what is today known as the battery was electrical science able to take a great leap forward: now there came to be available a source of constant, controllable, on-demand electric current. It was, according to Davy, a "key ... to lay open [...] nature." The story of how this invention came about begins with a grisly image: a severed tongue, writhing with unnatural animation under the impulse of eldritch energies.

THE TASTE OF IRON

Volta's dispute with Galvani on the source of the electricity causing the "galvanic response" (*see* page 43) prompted him to investigate in detail a phenomenon named contact potential. The German mathematician Johann Georg Sulzer first discovered this condition in 1752. He found that placing pieces of lead and silver in contact with one another on his tongue produced a distinctive taste, which he described as similar to "vitriol of iron" (iron sulphate). Clearly some sort of chemical reaction was occurring; Volta determined to repeat the experiment for himself.

He began with his own tongue: he covered the tip of it in tin foil, and then placed a silver coin near the root. When he pushed the coin forwards until it met the foil, he could clearly distinguish a strong sour taste. There was, however, no sign of twitching, so Volta moved on to a rather more grisly trial.

Cutting near the root, he severed the tongue of a freshly killed lamb and garnished it with tin foil over the cut surface. When he applied a silver spoon, Volta saw that the tongue trembled, while its tip began to move with ghastly animation. First it rose and folded back on itself, and then twitched from side to side.

These experiments proved to Volta that the contact between the two metals did indeed produce an electrical discharge, and – of paramount importance – he had experienced that the sour taste produced by the contact potential continued for the whole period in which the metals were on his tongue. Previous methods of generating electricity, such as electrostatic devices, had been capable only of producing momentary, irregular discharges.

Having established the principle, Volta now sought to discover which pairings of metal produced the greatest effect. Wiring combinations of metals to his ears and forehead, he managed to induce noises and bright flashes. Using an electrometer, a device for measuring electrical potential, Volta found that he was able to rank combinations of metals in order of their potentials relative to one another, with copper/silver and zinc proving to be particularly effective. (This is because copper and silver attract electrons strongly, while zinc releases them.) Crucially, he also deduced that the presence of some liquids, such as the saliva on his tongue, enabled the flow of current between the two metals. Liquids such as these are known as electrolytes (they contain charged particles known as ions that can carry electrical charge and are mobile).

VOLTA'S CROWNING ACHIEVEMENT

Meanwhile the animalists versus metalists dispute bubbled on. Galvani's supporters had performed more experiments in support of their animalist claims. In 1799, Volta responded by creating an entirely artificial model for demonstrating contact potential, with any organic components stripped out. In place of the tongue, which had provided the wet substrate for the current to flow between the two metals, he employed discs of cloth or cardboard soaked in salty water, placed between discs of silver (later versions used copper) and zinc. One such arrangement – or cell – produces only a tiny current, but the sum of the outputs of a whole series of them can be considerable. Accordingly, Volta constructed a pile of such discs, held in place by vertical glass rods: this was the first Voltaic pile.

Whereas earlier methods of generating electricity had produced discontinuous discharges – typically of high voltage but low current – Volta's pile generated a continuous current, with a low electrical potential but high current. Volta described electrical potential as "electrical tension", but it would later come to be described by the unit named after him: voltage. A common analogy is with water running through a pipe: voltage or potential is analogous to water pressure, while current (measured in amperes or amps) is analogous to flow (the volume flowing past in a given time).

In a pile, the cardboard discs tended to dry out over time, so Volta also constructed an alternative arrangement he called the *couronne de tasses* (crown or wreath of cups), in which cups of brine are connected in a ring by alternating metal strips of zinc and silver. A powerful current could be generated by a series or battery of these – powerful enough to enable entirely new kinds of chemistry.

A NEW ELECTRIC ORGAN

In a letter of March 1800, sent in two parts to the Royal Society in London – the premier scientific society of the age – Volta announced to the world what he described as an "artificial electric organ". He used it to bolster his case against Galvani's "animalists", suggesting that a similar "electromotive apparatus contained in the fish" explained the bioelectricity of the torpedo or electric ray, which he insisted to be "a simple physical, not physiological, phenomenon". "Not even in this case," he concluded, "is it proper to speak of animal electricity, in the sense of being produced or moved by a truly vital or organic action ..."

Volta seemed to be claiming that electric animals contain inorganic facsimiles of the battery or voltaic pile. However, this is a false premise, since the electric organ of the torpedo is composed of cells that generate electrical potential by pumping ions across membranes: very much a "physiological phenomenon". Nevertheless, there is a close resemblance between the gross structure of the natural electric organ and Volta's artificial one, since both consist of stacks of disc-like electric cells, which sum their potentials to create strong currents. In this sense Galvani and the animalists, with their electrophysiological model, were arguably more accurate, and it is they who surely had the greater influence on Shelley's conception of the power that animates Frankenstein's monster. However, Volta's critique of "vital action" added to a growing materialist chorus in this period; it was becoming ever easier to imagine there to be no absolute demarcation between organic and inorganic, vital and inanimate, dead matter and living tissue ...

In terms of the scientific *sequelae* of Volta's invention of the pile, its impact on the animalist/metalist debate was of secondary importance. Of far greater impact was the fashion in which this new technology opened a whole new field of science: electrochemistry. Before the second part of Volta's letter had even reached the Royal Society, British scientists Anthony Carlisle and William Nicholson had constructed their own pile and used it to "decompose" water: passing an electric current through water causes it to break down, where the water meets the electric probes, into its two constituent parts, hydrogen and oxygen. An important question in chemistry at the time was the identity of true elements – substances not capable of being "decomposed" into constituent parts – as opposed to compound substances. Decomposition by electric current, or electrolysis, promised radically to advance the task of distinguishing new elements. It was in this guise that Mary Shelley would have been given particular notice of the power and excitement of the new electrical technology, through meeting William Nicholson, a visitor to her father's London home, and through the research and lectures of Humphry Davy.

VOLTA AND NAPOLEON

Volta's pile won him fortune and glory. In 1801 he went to Paris to present his research to Napoleon himself, engaging him in the demonstration. The Emperor drew sparks from the pile, used its current to melt steel wire and even electrolysed water.

A BOUNDLESS PROSPECT

By 1800, Humphry Davy's ambition had moved beyond the Pneumatic Institute in Bristol, where he had been researching medicinal airs (*see* page 26). Immediately upon learning about it, Davy grasped the potential of the voltaic pile. His analysis of its action – showing that the pile generates current by the oxidization of zinc – won him election to the Royal Society and heralded his arrival on the grander stage of London's scientific community. In 1801, the Royal Institution (RI), a body for the pursuit and dissemination of scientific research, recruited Davy. Here he began a celebrated series of lectures that became immensely popular; attendees included one Mary Wollstonecraft Godwin.

The influential French chemist Antoine Lavoisier (1743–94) had predicted that the minerals known as soda and potash would prove to be oxides of hitherto undiscovered metals, if they could only be decomposed into their constituent elements. Davy was determined to discover the new metals, and he believed that the voltaic pile would enable him to do it. Accordingly, he lobbied the RI to raise money to build the most powerful pile yet constructed – with as many as 250 discs – and in 1807 he used it to electrolyse molten soda and potash, obtaining for the first time pure, elemental sodium and potassium. A famous account by his cousin, Edmund Davy, reveals some of the Romantic spirit that made Davy such an engaging and charismatic figure: "When he saw the minute globules of potassium burst through the crust of potash, and take fire as they entered the atmosphere, he could not contain his joy – he actually bounded about the room in ecstatic delight."

A year earlier, in a lecture at the RI, Davy had rhapsodized about the impact the pile would have: "a boundless prospect of novelty in science; a country unexplored", promising experiments that "can hardly fail to enlighten our philosophical system of the earth; and may possibly place new powers in our reach". Now he had fulfilled his own prophecies, cementing the status in the public imagination of electricity as a wonder-working technology, promising god-like powers to those who could master its mysteries. Although Shelley never actually specifies in the novel that Frankenstein is using electrical apparatus, ideas such as those so enthusiastically propounded by Davy must have informed her thinking, while electrical technologies would go on to become a mainstay of later adaptations of the *Frankenstein* story.

THE REVIVIFICATION MEN: GALVANIC REANIMATION AND ELECTRICAL THEATRE

"Perhaps a corpse would be re-animated," Mary Shelley recalled thinking in her preface to the 1831 edition of *Frankenstein*, "Galvanism had given token of such things." Indeed it had, for Galvanic revivification was the sharpest edge of contemporary medical research at the time of *Frankenstein*'s composition; demonstrations of the technique, fascinating and appalling in equal measure, had come to wide public attention, and these would become intimately bound up with Shelley's novel, which both took inspiration from the practice and provided it.

FOWL PLAY

Experiments such as those discussed by Benjamin Franklin (*see* page 38), allied to personal experience of the nasty shocks that Leyden jars were capable of delivering, prompted early speculation about the physical effects of electrocution, and the prospects that electrical shocks could undo what they had done. In 1755, Giovanni Bianchi experimented on dogs, using electrical shocks to induce seizures and cause the animals to stop breathing and then, he claimed, to revive them. This was a prelude to the first recorded case of deliberate defibrillation (the medical procedure in which electrical shock is used to restart a heart that has stopped beating), an experiment reported by the Danish veterinarian P. C. Abildgaard. According to his account published in the 1775 *Proceedings of the Medical Society of Copenhagen*:

> With a shock to the head, the animal was rendered lifeless and arose with a second shock to the chest; however, after the experiment was repeated rather often, the hen was completely stunned, and walked with some difficulty, and did not eat for a day and night; then later is very well and even laid an egg.

On the face of it, this report reads very much like a successful application of defibrillation to revive a chicken that had been killed by induced fibrillation (where an electrical shock stops the heart from beating). In practice, however, it is unlikely that a Leyden jar could supply enough charge for long enough for external defibrillation to have any effect, so Abildgaard's claims are suspect. However, the notion of electrical revivification was now current.

SHOCKS TO THE ANUS

There were many other similar experiments and claims, which spread the idea of electrical revivification. Dutch-born Swiss mathematician Daniel Bernoulli claimed to have revived drowned birds using sparks from an electrostatic generator. In 1796, Prussian naturalist Alexander von Humboldt inserted electrodes into the beak and anus of an unconscious bird and passed an electrical current through them, stimulating a rude awakening in the insensible beast. Similar methods were tried out on human subjects; along with electrodes in the mouth and anus, electrical current was applied to patients sitting in metal bathtubs filled with brine and bound with metal chains.

In London in 1774, the Royal Humane Society reported a celebrated case of apparent human resuscitation by electrical shock. A girl named Sophia Greenhill was the subject:

> A child three years old, fell from a one-pair-of-stairs window, upon the pavement, and was taken up without any signs of life. An apothecary being sent for, he declared that nothing could be done, and that the child was irrecoverably dead; but a gentleman who lived opposite to the place, proposing a trial with Electricity, the parents consented. At least twenty minutes elapsed before he could apply the shock, which he gave to various parts of the body without any appearance of success. At length, on sending a few shocks through the chest, a small pulsation became perceptible; soon after the child began to sigh, and to breathe, though with great difficulty: in about ten minutes, she vomited. A kind of stupor remained for some days; but she was restored to perfect health and spirits in about a week.

This case, too, has been cited as the first recorded instance of successful defibrillation, and was followed by others. In 1787, for instance, the Royal Humane Society reported a very similar case, in which a two-year-old boy, pronounced dead after falling out of a window, was revived by electrical shocks. In 1788, inventor Charles Kite wrote for the Society an "Essay on the Recovery of the Apparently Dead", in which he described a Leyden jar-style device that could generate up to 50,000V and produce a spark of 3cm (1.2 in). The Kite device was cited in a 1792 review of resuscitation cases by British scientist James Curry, which sets out protocols extremely similar to those used in modern defibrillation procedures. Despite all this, however, as with the Abildgaard claims, it seems unlikely that the available devices could have produced effective delivery of electrical shocks to the heart.

Nonetheless, in 1802 the Royal Humane Society published a report hyping the potential of electrical resuscitation, and proposing that electric shocks could be used as the definitive test for proving irrecoverable mortality (a major concern at a time when medical advances were rapidly shifting the threshold – see page 137).

ACTION ON THE RECTUM

Taking place precisely at this time were some of the most notorious experiments in electrical revivification, ghoulish demonstrations that may have directly inspired a key scene in Shelley's narrative. The man at the centre of these extraordinary public displays was Giovanni Aldini, nephew of Luigi Galvani, who in 1798 had become professor of physics at the University of Bologna, having taken up the baton of his illustrious relative.

Aldini continued to research "animal electricity" and its role in the contraction of muscles. Obtaining the carcass of a freshly slaughtered ox, he cut off its head and used an electrical current to twist its tongue. Turning to the decapitated body, he passed a high voltage through its diaphragm, with explosive results. The shock, he recorded, produced "a very strong action on the rectum, which even produced an expulsion of the faeces".

Since the passage of the 1751 Murder Act in England, it had become legal to use the bodies of executed murderers for experiments – partly in response to the shortage of medical cadavers and the unfortunate consequences of the growing demand (see page 128). In 1803, Aldini came to London to give a public demonstration of his research at the Royal College of Surgeons. Accounts of the dramatic results closely match key passages in the text of Frankenstein.

Galvani's nephew Aldini advanced his experiments, producing startling effects with ox heads – and if large mammals, why not man?

ON THE EVE OF BEING RESTORED TO LIFE

Aldini himself described the results of his trials on "the body of a malefactor executed at Newgate":

> *The first of these decapitated criminals being conveyed to the apartment provided for my experiments, in the neighbourhood of the place of execution, the head was first subjected to the Galvanic action. For this purpose I had constructed a pile consisting of a hundred pieces of silver and zinc. Having moistened the inside of the ears with salt water, I formed an arc with two metallic wires, which, proceeding from the two ears, were applied, one to the summit and the other to the bottom of the pile. When this communication was established, I observed strong contractions in the muscles of the face, which were contorted in so irregular a manner that they exhibited the appearance of the most horrid*

grimaces. The action of the eyelids was exceedingly striking, though
less sensible in the human head than in that of an ox.

A contemporary report from the *Times* of London gives more details, including the identity of the hanged man, one George Forster, describing how his body "was subjected to the Galvanic Process, by Professor ALDINI ..." It also includes a more sensational account of how Aldini "shewed the eminent and superior powers of Galvanism to be far beyond any other stimulant in nature":

> *On the first application of the process to the face, the jaw of the*
> *deceased criminal began to quiver, and the adjoining muscles*
> *were horribly contorted, and one eye was actually opened. In*
> *the subsequent part of the process, the right hand was raised and*
> *clenched, and the legs and thighs were set in motion. It appeared to*
> *the uninformed part of the bystanders as if the wretched man was*
> *on the eve of being restored to life. This, however, was impossible ...*

Aldini made gestures towards a respectable scientific rationale for such demonstrations, explaining that his experiment "was of a better use and tendency" and that Galvanism in general "offers most encouraging prospects for the benefit of mankind". In fact, Aldini was the first person recorded as having performed electro-convulsive therapy, or some prototypic form of it – administering electric currents to the heads of mentally ill patients and supposedly achieving remarkable results.

But there was also much of the showman about Aldini, as he regaled his audience with boasts of how he had made the hand of a headless man clutch a coin and throw it across a room, or related a tale of how he had wired up the heads of two decapitated criminals arranged to face one another, galvanizing them into such awful grimaces "that the Spectators ... were actually frightened". Indeed, according to one account, Aldini's London demonstration was so terrifying that a beadle attending the show died of fright!

Such lurid and dramatic effects cannot have been lost on any aspiring writer drawing inspiration from accounts like these, and it seems plausible that there is a direct link between the grotesque contortions of Forster's face, during which "one eye was actually opened", and Shelley's description of the monster's awakening: "I saw the dull yellow eye of the creature open; it breathed hard, and a convulsive motion agitated its limbs."

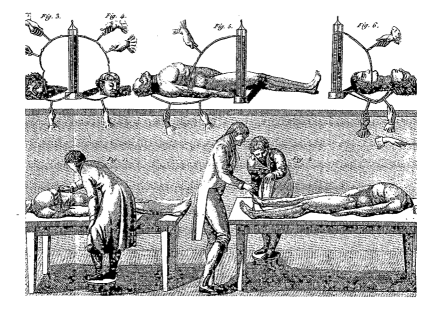

*A macabre series of galvanic experiments on corpses,
severed heads and combinations of the two.*

A GOTHICK EXPERIMENT

Could the line of transmission – between Galvanic revivification experiments and Shelley's *Frankenstein* – have run both ways? In 1818, the very year of the novel's publication, Scottish chemist Andrew Ure attempted a demonstration more ambitious than any staged by Aldini. Theorizing that previous efforts had been hampered by underpowered voltaic piles, and because shocks were administered to the mass of muscle tissue rather than to the nerves innervating the muscles, Ure determined to prove, at least in principle, that Galvanic revivification was viable: a dead body *could* be given life. As Danish literary academic Ulf Houe, writing in the journal *Studies in Romanticism*, observes, "While Aldini contented himself with the role of spasmodic puppeteer, Ure's ambitions were well nigh Frankensteinian."

Ure's demonstration took place in November, in the anatomical dissection theatre at the University of Glasgow. The corpse of an athletic young murderer named Clydesdale was brought to the theatre "about ten minutes after he was cut down", Ure details in his account of the experiments. Five minutes before this, Ure had charged with nitric and sulphuric acid a powerful battery, "consisting of 270 pairs of four inch plates" connected to "pointed metallic rods with insulating handles, for more commodious application of the electric power". Incisions were made into the body to expose nerves at the neck, hip and heels, and these were then shocked with the rods, with remarkable results.

Applying the rods to the heel, the corpse straightened its bent leg with such force as to send one of the assistants flying. When the sciatic nerve was stimulated, "every muscle of the body was immediately agitated with convulsive movements, resembling a violent shuddering from the cold." With the rods applied to a cut in the tip of the forefinger, "the fist being previously clenched, that finger extended instantly; and from the convulsive agitation of the arm, he seemed to point to the different spectators, some of whom thought he had come to life".

The demonstration grew more macabre still. On stimulating Clydesdale's forehead:

> Every muscle in his countenance was simultaneously thrown into fearful action; rage, horror, despair, anguish, and ghastly smiles, united their hideous expression in the murderer's face, surpassing far the wildest representations of a Fuseli [romantic painter] or a Kean [tragic actor]. At this period several of the spectators were forced to leave the apartment from terror or sickness, and one gentleman fainted.

But the result that excited Ure the most came when he passed electricity through Clydesdale's diaphragm: "The success of it was truly wonderful. Full, nay, laborious breathing instantly commenced. The chest heaved, and fell ..." Ure was convinced that if the body had not been that of a hanged man who had suffered severe bodily injury, and been drained of blood, he could have restored life to the corpse.

A GALVANISED CORPSE

Cartoon illustrating contemporary anxieties about the prospect of galvanic reanimation; just a short step from Frankenstein in his laboratory.

Ure's claims seem to have met with little approbation. On the contrary, a pamphlet he wrote about the demonstration was slated as "publicity of the crudest kind" by W. V. Farrar, writing in *Notes and Records of the Royal Society of London,* who dismissed the whole affair as "This rather 'Gothick' experiment, reported in such appropriate literary style ..."

Mary Shelley's novel achieved its earliest fame through numerous popular stage adaptations, the first of which appeared within a few years of the book's publication. In this incarnation, *Frankenstein* thus reflects the theatrical nature of the demonstrations that helped to inspire it. Indeed for all its academic significance, the electrical science of the era had something of a disreputable air, with public demonstrations often carried out in rooms in public houses – as much shows of popular entertainment as educational events. Was Ure's stunt perhaps directly influenced by the publication of *Frankenstein*, eleven months earlier, making this, in some sense, the original theatrical adaptation of the novel?

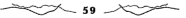

CHAPTER 3

THE RIGHT STUFF

VITALISM, SPONTANEOUS GENERATION AND THE MEANING OF "LIFE"

AS THE SCIENTIFIC REVOLUTION EXTENDED INTO BIOLOGY, SO THE FIELD BECAME RIVEN BY FIERCE DEBATES OVER THE VERY MEANING OF LIFE ITSELF AND THE NATURE OF THE VITAL FORCE. ANCIENT SUPERSTITIONS ABOUT SPONTANEOUS GENERATION GAVE WAY TO NEW HYPOTHESES OF SEEDS AND SPORES. THE POSSIBILITY OF SCIENTIFIC MASTERY OF THE VITAL SPARK ITSELF EMERGED, ALBEIT DIMLY, INTO VIEW.

THE QUICK AND THE DEAD:
WHAT GIVES LIFE?

"I collected the instruments of life around me, that I might infuse a spark of being into the lifeless thing that lay at my feet." The animation of the creature is one of the pivotal events in modern literature, and particularly in the history of science fiction, yet it takes just a handful of words: Frankenstein "[infuses] a spark of being into the lifeless thing". The brevity of this description allows Shelley to evade a host of objections and assumptions, and is seen as one of the many elements that give her creation its timeless and universal power (see below). Yet her phrasing, be it ever so simple, is freighted with meaning, alluding to one of the great debates of her age – a debate in which the novel would come to play a central, though heavily contested role. What is life? What is the difference between the living and the dead? Is there an animating force or principle, a "spark of being"? Would the answer to such questions be changed if it were indeed possible to reanimate dead tissue and bring life to the lifeless? Thus Shelley's novel goes straight to the heart of the debate between vitalism and mechanism.

FIRST PRINCIPLES

Vitalism is the doctrine that life cannot be explained in or reduced to purely physical or chemical terms, and that living things possess some sort of special principle, whether it be soul or life force. The opposing doctrine, mechanism (also sometimes described as materialism or physicalism, although these are slightly different), holds that life can be explained in purely physical or chemical terms – that there is no fundamental difference between animate and inanimate matter, and no such thing as a vital spirit. The debate between these two opposing schools of thought was extremely current in the Shelleys' circle at precisely the time at which Mary was gestating her novel; indeed she recounts that among the "various philosophical doctrines" she discussed with Byron and Shelley was "the nature of the principle of life, and whether there was any probability of its ever being discovered and communicated". It was a debate that reflected major developments in the theory and practice of biology. Vitalism itself, however, had antecedents stretching back at least as far as Aristotle.

ARISTOTLE'S ENTELECHY

Simple observation remains one of the most powerful arguments for vitalist thinking. At the moment of death, an animal is both utterly transformed yet exactly the same; though life has fled, the breast is stilled and the heart stopped, the matter that comprises a creature appears unchanged. The same assemblage of tissues at one moment possesses life – with all that this entails – and at the next moment it is merely meat. Perhaps the most natural conclusion to be drawn from such observations is that the difference between the two states is some sort of vital force or principle, a spirit or soul.

Aristotle, the most influential natural philosopher until the Renaissance, and a pioneer of the study of biology and natural history.

This was the conclusion that Aristotle drew from his biological studies. He argued that a living organism is distinguished from the mere matter of which it is comprised by a soul or vital force that he termed its "entelechy": a guiding principle that makes actual what is merely potential. Thus, the vital force not only gives life but actually guides development, and this helped to solve a related problem: how do living things develop form and organization out of formless, undifferentiated matter – that is, how does the egg become a chicken?

Not all the Classical philosophers endorsed a vitalist position. Atomists and Epicureans, such as Thales, Democritus, Epicurus and (later) Lucretius, were materialists who saw matter and motion as the essential principles, downplaying the role of the immaterial and spiritual. However, the Graeco-Roman physician Galen did espouse vitalism, holding that vital spirits are necessary for life – and since he and Aristotle were the most powerful influences on medieval Western biology, vitalism was the default position until the Enlightenment. Aristotle's entelechy – with its implications of a guiding purpose to animation and development – fitted into a religious worldview. God infused living beings with their vital sparks, or souls, and this in turn provided the organizing principle by which natural creation could be understood.

MEAT MACHINES

What seemed like a medieval vitalist consensus began to come apart in the Enlightenment, as enquiring minds delved increasingly deeper into the anatomy and chemistry of organic matter. Among the most influential voices was French philosopher René Descartes, who made a detailed study of anatomy. According to one legend, when asked to show a visitor his library, Descartes led him to the dissecting room and pointed to the dissected specimen on the table, declaring, "There is my library!"

Descartes' readings in this library of living matter led him to view living tissue, up to and including the human body, in increasingly mechanistic terms. Teasing out tendons and muscle fibres, Descartes was put in mind of the strings and pulleys of a machine: "I do not recognize any difference between the machines made by artisans and the various bodies which nature alone constructs." The only exception he made was to allow that while the machines of artisans must inevitably be at a scale suited to the hands that constructed them, "the tubes and springs which cause the effects of natural bodies are ordinarily too small to be seen".

Following this vision of nature as machine, a doctrine opposing vitalism came to be known as mechanistic; but in fact Descartes did not discard vitalism by any means. He specifically referred to "animal spirits" as the animating principle of the meat machines he observed, although admittedly this was more a variation on his mechanistic theme that was sometimes called the hydraulic theory of neuromuscular action (*see* page 40). More relevant to the vitalist question, Descartes insisted that man was not merely machine, and that humanity derived from an animating soul. Thus, he removed the question of the vital spirit to the realm of the philosophy of consciousness, where his dualism (belief in the essentially different natures of mind and body) stood in opposition to materialism or physicalism (belief that mind can be explained solely by reference to material or physical facts). This debate is yet another of the themes explored by Mary Shelley's novel, and is discussed in Chapter 4.

LITTLE LIVES

What of non-human animals, though, and of the organs, tissues and matter of the body? Here Descartes did appear to be striking at the notion of vitalism, with its clear distinction between animate and inanimate. The mechanistic characteristics of his philosophy came to be attributed to the whole body of French Enlightenment natural philosophy, although in practice it was extremely diverse. The eighteenth-century French naturalist George-Louis Leclerc, Comte de Buffon (1707–88), was an influential voice in the vitalist camp, drawing a clear dividing line between *matière vive* ("living matter") and *matière brut* ("inanimate matter"). Buffon described living bodies as being constructed from particles of the former, distinguished by its tendency towards activity and motion, with life arising from the collective action of such "molecules":

> *The life of the whole (animal or vegetable) would seem only to be*
> *the result of all the actions, all the separate little lives, if I may be*
> *permitted so to express myself, of each of those active molecules*
> *whose life is primitive and apparently indestructible.*

Buffon's molecules of *matière vive* were typical of vitalist theories. Gottfried Wilhelm Leibniz, German polymath and rival to Sir Isaac Newton, spoke of monads, while French philosopher Pierre Louis Moreau de Maupertuis discussed corpuscles, arguing that increasing complexity of organization of such bodies leads to the emergence of intelligence. French encyclopaedist Denis Diderot described "living points" or "living molecules".

SENSIBLE CARBON

The vitalist position became increasingly complex, and arguably confused. It embraced religious theories ascribing supernatural origins to the vital spark, metaphysical but not necessarily religious approaches, and atheistic and even radically materialist views. Among the schools of thought most influential on the Shelleys was the German Romantic *Naturphilosophie* of Friedrich Schelling and the circle of thinkers and writers centred on the University of Jena. Described as "scientific mysticism", *Naturphilosophie* imagined a kind of entelechtical animism infusing the world and driving evolutionary progression, so that all forms of matter, from rocks and stones up to plants, animals and humans, are imbued with spiritual energy that makes them "aspire" to higher states. While *Naturphilosophie* helped inspire great writing and breakthrough science – as with Johann Ritter, who was led to discover ultraviolet light (*see* page 168) – it could also become ridiculous. In one infamous example, Scandinavian geologist Henrick Steffens proclaimed that: "a diamond is a piece of carbon that has come to its senses".

At the other end of the spectrum were apparent materialists (those who reject the role of immaterial spirit), such as the French anatomist Xavier Bichat. Bichat identified twenty-one distinct types of tissue that made up living systems, and categorized them according to their vital properties, which varied along dimensions of "sensibility" and "contractility". He described these as fundamental forces of nature on a par with Newton's gravity and "endowed by God", and argued that through their action living tissue resists decomposition. This led to a decidedly pared-down form of vitalism, in which he defined life as "the sum of the functions by which death is resisted".

THE SUBTLE SUBSTANCE

British vitalism tended to be more considered and less radical. The eminent surgeon and anatomist John Hunter essentially aped Buffon's theory, but in Latin, changing *matière vive* to *materia vitae*. His successor John Abernethy, who became Professor of Anatomy at the Royal College of Surgeons, developed this further into a semi-mystical "Vitality": "a subtle, mobile invisible substance, super-added to the evident structure of muscles, or other form of vegetable or animal matter, as magnetism is to iron, and as electricity is to various substances with which it may be connected".

Abernethy's conception clearly left space for God – something must have "super-added" the subtle substance – and mystical forms of vitalism such as this, and the wild speculations of the Naturphilosophists, provoked a kind of materialist backlash in London, with the 1793 lectures on "Animal Vitality" by the radical John Thelwall. Thelwall rejected any role for the supernatural as the source of the spark of life, insisting that: "Spirit, however refined must be material." This apparently self-contradictory statement stimulated lively debate, especially in one influential set – the young Romantic poets, such as Wordsworth and Coleridge. They would, in turn, transmit to the Shelleys a kind of ambivalence about vitalism, which contested with a straightforward rejection of vitalism by Percy Shelley's doctor and mentor in biological science, William Lawrence.

CHEMICAL VITALISM

As new branches of science opened up, each seemed to offer new hope of tracking down the elusive vital spirit. In chemistry oxygen was isolated from "inferior air" and identified as essential to life (*see* page 22). The French chemist Antoine Lavoisier, with his colleague Pierre-Simon Laplace, demonstrated that respiration in animals (the process by which food is converted into energy and waste chemicals) is a form of "slow combustion"; chemical analysis was revealing the biochemistry of life. As organic chemicals were identified and differentiated from inorganic ones, so it appeared that chemistry might reveal the secret of vitality. Eventually chemists began to realize that there is no clear distinction between organic chemicals produced in living things and those that the chemist could create in the laboratory. On the one hand this struck at the roots of vitalism, with the dominant chemist of his generation, the Swede Jacob Berzelius, writing in 1836 that "There is no special force exclusively the property of living matter which may be called a vital force". On the other hand, such advances added to the growing conviction that life is not utterly distinct from inanimate matter, and that science might soon show how one could easily be transformed into the other.

Lavoisier at work in the laboratory; studies such as the one shown here, on respiration, offered the prospect of a "chemical vitalism".

ELECTRIC VITALISM

The most obvious candidate for the elusive vital spark was electricity, and the experiments of Galvani, Aldini *et al* (*see* Chapter 2) appeared to demonstrate a conclusive link: that electricity could produce muscular contraction, bodily movement, and breathing and even restore life (*see* pages 50–9). Adam Walker, a friend of Joseph Priestley, breathlessly averred: "Its power of exciting muscular motion in apparently dead animals, as well as of increasing the growth, invigorating the stamina, and reviving diseased vegetation, proves its relationship or affinity to the living principle ... it is impossible not to believe it is the soul of the material world ..."

FROM AN OYSTER TO A MAN

William Lawrence (1783–1867) had been Abernethy's protegé and successor as Professor of Anatomy at the Royal College of Surgeons. However, unlike his dourly pious and conservative mentor, Lawrence had eagerly imbibed many of the new-fangled continental philosophies. As a student he had studied at Göttingen University in Germany under the prominent anatomist and early anthropologist Johann Friedrich Blumenbach, and translated into English Blumenbach's *Comparative Anatomy* in 1807. The German had amassed, measured and classified a vast collection of skulls, and his work had greatly advanced the case for a link between craniology – the physical structure of the brain – and the production of mind. It was a controversial and radically materialist position, with Blumenbach coming dangerously close to rejecting the existence of the soul. Lawrence was also familiar with

Bichat's bleak quasi-mechanistic theories (*see* page 66), and with the work of radical French physiologist Julien Offray de La Mettrie, whose "Machine Man" theories adopted a hard-line mechanistic approach to physiology, in which humans were viewed as nothing more than "perpendicularly crawling machines" (*see* page 111).

Commencing in the spring of 1816, Lawrence would introduce this radical materialism to an English audience in his Hunterian Lectures – the prestigious public lecture series that was begun by Hunter and continued by Abernethy. To the astonishment of many, Lawrence used the occasion not to praise his forebears but to bury them, launching a ferocious assault on vitalism. He utterly dismissed the "Life Principle" of Hunter and Abernethy, insisting that the human body is merely a complex physical organization, and famously describing how the development of this organization could be traced throughout the animal kingdom, "from an oyster to a man".

Lawrence was scathing about the way in which the defenders of vitalism constantly shifted the locus of the vital force: "To make the matter more intelligible, this vital principle is compared to magnetism, to electricity, and to galvanism; or it is roundly stated to be oxygen. 'Tis like a camel, or a whale, or like what you please ..." He denied any role for theology or metaphysics in what he cast as a purely scientific issue: "The theological doctrine of the soul, and its separate existence has nothing to do with this physiological question ... An immaterial and spiritual being could not have been discovered amid the blood and filth of the dissecting room."

In summary, Lawrence rejected vitalism as unscientific myth making:

> *It seems to me that this hypothesis or fiction of a subtle invisible matter, animating the visible textures of animal bodies, and directing their motions, is only an example of that propensity in the human mind, which had led men at all times to account for those phenomena, of which the causes are not obvious, by the mysterious aid of higher and imaginary beings.*

THE UNKNOWABLE FACTOR

Vitalism is of interest here because of its undoubted influence on Mary Shelley's thinking and the writing of *Frankenstein*, and it continued to be a hot topic in biology until well into the twentieth century. However, today it has vanished completely from scientific discourse. One reason for its disappearance, as discussed above, is because vitalism arose as a solution to what seemed like an otherwise intractable problem; as an explanation for apparently irreducible complexity. How else could a spot of undifferentiated protoplasm give rise to a fully-grown animal? How else to explain the million different properties of the living organism? Advances in biology, particularly in molecular genetics, have more or less dispelled this problem, so that vitalism is no longer a useful concept. But vitalism also suffers from a more fundamental flaw: it is a fallacy, a tautology of circular logic that seeks to explain through reference to itself. Living things are alive because they are made of vital particles. Life is defined by the presence of vital force, which is that property that makes life unique. German-American evolutionary biologist Ernst Mayr said that vitalism "virtually leaves the realm of science by falling back on an unknown and presumably unknowable factor". In particular, vitalism has been accused of being unfalsifiable, because it offers no testable predictions. German-American philosopher C. G. Hempel complained that vitalistic explanations "render all statements about entelechies inaccessible to empirical test and thus devoid of empirical meaning" (in other words, experiments cannot be done on them).

MURDER TO DISSECT

Lawrence's attack on vitalism aroused a strong reaction. For example, the conservative *Quarterly Review* complained that Lawrence asked readers to believe "that there is no difference between a man and an oyster, other than that one possesses bodily organs more fully developed than the other! That all the eminent powers of reason, reflexion, imagination, and memory ... are merely the function of a few ounces of organized matter called the brain!" More dangerously still, Lawrence's discourse had clear overtones of atheism, and it – and the vitalist debate in general – soon became subsumed into a spurious dichotomy between brutally reductive, atheistic and predominantly French science, and a humane, religious English version.

English Romantic poetry found itself dragooned into this battle on the latter side, thanks to the quickly mythologized Immortal Dinner of 1817 (*see* box opposite), at which John Keats was said to have blamed Newton for "unweaving the rainbow" (in fact, Keats actually said that Newton had "destroyed all the poetry of the rainbow" – "unweaving the rainbow" is a line from his later poem "Lamia"). This reference harked back to a much earlier line from Wordsworth's poem "The Tables Turned", which accused "meddling intellect" of misshaping "the beauteous forms of things: We murder to dissect."

Samuel Taylor Coleridge and Percy Shelley – perhaps better informed than most of those who attended the dinner, and certainly more supportive of science – rejected this supposed dichotomy between poetry and science. Coleridge entered the vitalism debate, attempting to thread a middle course. He denied that life was purely physical organization, but he also wrote: "I must reject fluids and ethers of all kinds, magnetical, electrical, and universal, to whatever quintessential thinness they may be treble-distilled ..." Coleridge (like Descartes) removed the location of the vital force from simple physiology to the related issue of consciousness, shifting the debate into the field of philosophy of consciousness and the contest between dualism and materialism (*see* page 109). His approach thus anticipated the eventual fate of vitalism, which was gradually dispelled by increasing knowledge of the mechanisms and possibilities of physiology, development and

THE IMMORTAL DINNER

Benjamin Haydon was a celebrated painter, devoted Christian and society figure, who was friendly with several Romantic poets, including Wordsworth, Charles Lamb and John Keats. In December 1817 he invited all three of them, along with various others, to a luncheon at his house – partly to introduce Keats to Wordsworth, and partly to celebrate getting halfway through his monumental composition *Christ's Entry into Jerusalem*, a huge painting that presided over the meal. Conversation turned to the inclusion of various figures into the crowd observing Jesus, which included both Wordsworth and Keats, and also other important Enlightenment names such as Voltaire and Newton. Becoming rather garrulous, Lamb began to mock Voltaire, before moving on to Newton. Haydon recounted the scene in his *Autobiography*:

[Lamb] then, in a strain of humour beyond description, abused me for putting Newton's head into my picture; "a fellow," said he, "who believed nothing unless it was as clear as the three sides of a triangle". And then he and Keats agreed he had destroyed all the poetry of the rainbow by reducing it to the prismatic colours. It was impossible to resist him, and we all drank "Newton's health, and confusion to mathematics."

biology in general. Vitalism had arisen primarily as a solution to the problem of how to explain life and particularly of how to explain how complex organisms developed from unorganized and apparently simple matter, which was not evidently different from inanimate matter. Understanding of genetics, DNA, protein manufacture, inter-cellular signalling, and so on, has effectively dispelled this problem, but left in place the more intractable problem of consciousness.

A MOMENT OF EQUIVALENCE

Mary Godwin was exposed to the fluctuating currents of thought about vitalism through several different avenues: her own wide reading; her experience of the intellectual circle centred on her father; her attendance at medical demonstrations and scientific lectures; her lover and soon-to-be-husband Percy Shelley. Percy was a patient and confidant of William Lawrence, and their medical consultations came at the height of the controversy over vitalism. Their conversations extended from medical complaints to literary and scientific digressions, and Lawrence almost certainly helped educate Percy and Mary about aspects of French and German thinking, for it is likely that Mary occasionally accompanied Percy to the consultations.

In 1814, the young couple travelled to the Continent, as they were forced to elope when Mary's father forbade her to see Percy. Here, their journal records them discussing aspects of vitalism. In Switzerland, in 1817, Mary listened to Percy and Byron discussing the controversy – with the physician John Polidori on hand to offer, presumably, another informed view. Did Mary embrace Lawrence's attack on vitalism, or did she reject it? What does the novel have to say about vitalism? Does it come down on one side of the debate?

Well-informed early readers might have assumed that the anatomical themes of the novel aligned its author with Lawrence's view. Additionally, the basic facts of the tale plainly speak to a materialist position: a scientist is able to bring inanimate matter to life through purely physical methods. However, in reality both the text and readings of it have been much more ambiguous, nuanced and equivocal. For example, leading Romantic scholar Marilyn Butler argued that although Victor Frankenstein was like Abernethy – "super-adding" a "spark of being" to dead matter to bring it to life – the novel itself takes a position more akin to that of Lawrence, commenting archly on Victor's "serio-comic" blundering. English professor Janis Caldwell points out that from the earliest stage adaptations to Mary Shelley's 1831 revision to the later versions, "the popular imagination interpreted Frankenstein as a cautionary tale warning against the presumptions of a purely materialist science".

In fact, the novel equivocates between vitalist and materialist positions; Caldwell says that "Shelley seems to hover between philosophical positions". Francis Bacon, the pioneering early modern philosopher of science, cautioned against "unwise mixing" of the two books of philosophy, the theological and the natural, and this was echoed by Lawrence when he counselled against seeking the spiritual being in the blood and filth of the dissecting room. Yet Mary is unabashedly guilty of this unwise mixing, situating the quest to create a spiritual being within the dissecting room, co-mingling the material with the transcendental. This mixing extends to her account of the monster's mental development, which combines materialistic ideas, about sensations impressed on nervous tissue, with transcendental language describing the innate goodness of the natural child. Even the monster himself seems confused, referring both to the ephemeral materiality of his body – "my ashes will be swept into the sea by the winds" – and to the possibility that he has an immortal soul: "My spirit will sleep in peace".

The key to the novel's ambiguity lies in that masterfully brief creation scene – and in Shelley's refusal to define the method of the monster's animation, her rejection of a simple choice between materialism and transcendental vitalism. Martin Willis, a specialist in nineteenth-century science fiction, argues:

> In leaving the creation of the monster equivocal, the vital turning point between inertia and animation can be appropriated by either materialist or Romantic science. Moreover, without either of these opposing philosophies able to defend their position from textual evidence, they both exist simultaneously, caught in a moment of equivalence at the very centre of the novel.

ANIMATED PASTA: SPONTANEOUS GENERATION AND THE ORIGINS OF LIFE

One of the most peculiar lines in Mary Shelley's preface to the 1831 edition of *Frankenstein* appears to refer to a piece of spaghetti coming to life. Describing some of the topics discussed by Percy Shelley and Lord Byron, Mary recalls:

> *They talked of the experiments of Dr Darwin (I speak not of what the Doctor really did, or said that he did, but, as more to my purpose, of what was then spoken of as having been done by him) who preserved a piece of vermicelli in a glass case, till by some extraordinary means it began to move with voluntary motion. Not thus, after all, would life be given.*

Mary herself raises caveats about the account, first by making it clear that the actual work of Dr Darwin did not necessarily bear any relation to what was discussed ("I speak not of what the Doctor really did"), and secondly by not unreasonably dismissing the notion that the spontaneous animation of a piece of pasta might unlock the secret of life. Yet the passage is extraordinary nonetheless. Who is "Dr Darwin", and why is he spoken of as having brought to life a piece of vermicelli?

LIFE FROM NON-LIFE

The good doctor in question is Erasmus Darwin, the grandfather of Charles Darwin and one of the pre-eminent figures in eighteenth-century science, particularly in biology. Erasmus Darwin was a physician, poet, naturalist, inventor and botanist, who prefigured many of the themes that would be explored by his grandson, including evolution and the origin of species. But the reference in Mary's preface relates to another biological doctrine: spontaneous generation. Also sometimes known as "equivocal generation", spontaneous generation is the belief that new generations of organism can arise not through sexual reproduction, but from non-living matter, whether decaying organic matter (heterogenesis) or inorganic matter (abiogenesis).

Spontaneous generation was an important influence on the inception of Shelley's novel, both directly through Mary's familiarity with the concept, and more generally as part of the intellectual background of the era; the underlying preconceptions and assumptions that helped to make Frankenstein's monstrous creation imaginable. The doctrine of spontaneous generation reflected a fundamental belief in the potential for the inanimate to become animate, especially through the application of an animating principle, or in Mary Shelley's words, when it was "imbued with a spark of being". The existence and nature of this spark is discussed elsewhere (*see* page 62), but the very possibility of its being effective depended on a conception of the phenomenon of spontaenous generation. More specifically, spontaneous generation was understood since classical times to be a particular property of decaying and putrid matter, so that Victor's charnel-house gleanings – the corpses and body parts he gathers – are qualified for generation by the very properties that, to modern eyes, make them unsuitable for reanimation.

BOUGONIA

The first century BCE Roman poet Virgil detailed a kind of spontaneous generation technology named *bougonia* ("ox progeny") – a traditional belief that bees could be generated from the carcasses of cattle. He even gave instructions for the process in his *Georgics*:

> *Then seek they from the herd a steer, whose horns*
> *With two years' growth are curling, and stop fast,*
> *Plunge madly as he may, the panting mouth*
> *And nostrils twain, and done with blows to death,*
> *Batter his flesh to pulp i' the hide yet whole,*
> *And shut the doors, and leave him there to lie.*
> *But 'neath his ribs they scatter broken boughs,*
> *With thyme and fresh-pulled cassias: this is done ...*
> *Meanwhile the juice within his softened bones*
> *Heats and ferments, and things of wondrous birth,*
> *Footless at first, anon with feet and wings,*
> *Swarm there and buzz, a marvel to behold.*

THE CHICKEN WITHOUT THE EGG

Spontaneous generation was received wisdom in both educated and popular understanding of biology; its roots went far back in time. The Bible features instances of life arising from inanimate matter, as when Aaron's staff turns into a snake, but a scientific gloss was applied to the doctrine through the authority of ancient Greek philosophers, beginning in the sixth century BCE with Anaximander of Miletus, who proposed that life could originate spontaneously by the action of light on water, and explained that fish grew from warm mud.

Anaximander's theory was extended by subsequent philosophers and elucidated to the highest degree by Aristotle, who again stressed the role of heat:

> *Animals and plants come into being in earth and in liquid because there is water in earth, and air in water, and in all air is vital heat so that in a sense all things are full of soul. Therefore living things form quickly whenever this air and vital heat are enclosed in anything.*

In works such as the *Generation of Animals*, Aristotle gave examples of types of spontaneous generation:

> *So with animals, some spring from parent animals according to their kind, whilst others grow spontaneously and not from kindred stock; and of these instances of spontaneous generation some come from putrefying earth or vegetable matter, as is the case with a number of insects, while others are spontaneously generated in the inside of animals out of the secretions of their several organs.*

Spontaneous generation was extended still further by the Roman proto-evolutionary theorist Titus Lucretius Carus in his poem *On the Nature of Things* (c. 50 BCE), which was a major influence on Erasmus Darwin (see page 83). According to Lucretius, while the present-day Earth is old and tired and thus limited to spontaneously generating lower forms of life such as worms, in the past all forms of creature were brought forth spontaneously, a scheme that explicitly extended to humans.

*Aaron's rod changes into a serpent – a Biblical
instance of apparent spontaneous generation, as life
arises from inanimate matter.*

The doctrine memorably featured in Ovid's *Metamorphosis* (8 CE),
which describes how, "When the sun's radiance warmed the pristine
moisture; And slime and ooze marshlands swelled with heat ... the
seeds of things ... gained life; And grew and gradually assumed a
shape." As Aristotle had outlined, animals were assigned specific
matrices of origin: bees from carcasses of cattle (*see* box on page 77),
wasps from dead donkeys or horses, scorpions from dead crabs and
snakes from rotting spinal cords.

WEAKER THAN NATURE
Several medieval Islamic scholars took up the spontaneous
generation doctrine. In the eleventh century the Islamic philosopher
Ibn Sina (known in the West as Avicenna) claimed that snakes could
spontaneously generate from women's hair incubated in a warm, wet
environment. Avicenna is also the author of a famous line relating

to artificial attempts to recreate nature, with obvious relevance to *Frankenstein*: "Artifice is weaker than nature and does not overtake it."

This theme was taken up by the twelfth-century Islamic scholar Ibn Rushd (known in the West as Averroes), who discussed Aristotle's *Generation of Animals* with specific reference to the difference between creatures spontaneously generated from putrid matter (including those artificially thus created) and those resulting from ordinary sexual reproduction. Averroes claimed that although superficially similar, such animals are actually different and, in some senses, inferior; they would, for instance, be sterile.

SWALLOWS IN THE MUD

Today spontaneous generation takes its place alongside other biological doctrines that were widely accepted from classical to early modern times, and yet which now seem like eccentric curios, such as beliefs about birds and to where they disappeared when they migrated. Lines from the *Iliad* – "cranes... who flee the winter and the terrible rains; and fly off to the world's end" – acknowledge that birds migrate to far-away destinations, yet Aristotle recorded that redstarts transmute into robins, apparently unaware that the one species simply replaces the other in winter, as redstarts migrate to Africa and robins migrate to Greece from further north. One of the most widespread beliefs was that geese develop from barnacles, accounting for their sudden appearance in spring, and in some parts of Europe Catholics took this to mean that the goose was fair game on "fish-only Fridays". Writing in 1187, Gerald of Wales recorded that: "in some parts of Ireland, bishops and men of religion make no scruple of eating these birds on fasting days, as not being flesh, because they are not born of flesh".

In the sixteenth century, cartographer Olaus Magnus promoted the theory that swallows disappear in winter because they dive to the bottoms of rivers and bury themselves en masse in the clay, in order to hibernate. In the seventeenth century, English minister and teacher Charles Morton went one better and claimed that birds migrate not to other continents but to the Moon. It was widely believed at the time that the Moon might harbour life and have an environment similar to that of Earth, but even so Morton's scheme called for birds to travel for 60 days at 125 mph in order to cover a distance he estimated to be 179,712 miles.

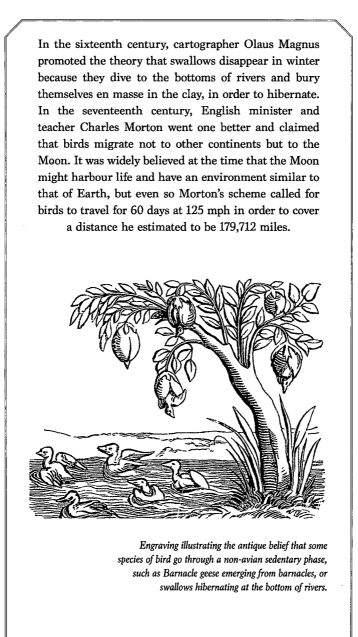

Engraving illustrating the antique belief that some species of bird go through a non-avian sedentary phase, such as Barnacle geese emerging from barnacles, or swallows hibernating at the bottom of rivers.

THE ORIGINS OF MAN

The spontaneous generation doctrine persisted in Europe as folk wisdom, and also passed on through the transmission of Islamic and Classical sources. Lucretius was especially influential on Renaissance and Enlightenment thought, and one particularly interesting manifestation of this was the propagation of his concept of the spontaneous generation of humans; in other words, the notion that a human might be created not by sexual reproduction from human parents, but from inanimate matter. As one of the most powerful exponents of atomism – the natural philosophy that everything is composed of indivisible particles, and the use of this as the basis for a largely materialist explanation of natural phenomena – Lucretius became the inspiration for a new generation of atomists. Atomism was updated and popularized by scholars such as the seventeenth-century French philosopher and mathematician Pierre Gassendi, and its materialist discourse meshed with the mechanistic philosophies of Descartes (see page 116) and the science of Newton and Boyle, to create a powerful and all-embracing synergy.

The new atomism thus propagated to such an extent that it produced a reactionary backlash from those who feared its materialist and atheistic implications. Ironically, it was the writings of critics of the new atomism, such as the English jurist Matthew Hale's 1677 *The Primitive Origination of Mankind, Considered and Examined According to the Light of Nature*, and Bishop Richard Bentley's 1692 series of sermons "A Confutation of Atheism from the Structure and Origin of Humane Bodies", which actually did the most to popularize the idea that perhaps humans could be created without parents, from inanimate matter. Hale and Bentley saw as a particular challenge the widespread assumption as fact that insects could be spontaneously generated. Everyone "knew" that bees and flies came from rotting meat, and that insects generated in this way appeared to be perfectly normal animals. By analogy, this cleared the way for the idea that humans too could have arisen – and might yet arise? – in a similar fashion. The door was open for the concept of a new type of anthropoeia (science writer Philip Ball's term for "making people"), in which science or artifice would replace nature.

THE VERMICELLI MISTAKE

One of Lucretius's greatest disciples was Erasmus Darwin, and he gave examples of spontaneous generation or related phenomena that may have influenced and/or inspired Mary Shelley's thinking. Amongst the copious footnotes to his epic 1802 poem *The Temple of Nature*, Darwin describes microscopic animals (also known as animalcules) named "vorticellae":

> *Thus the vorticella or wheel animal, which is found in rain water that has stood some days in leaden gutters ... though it discovers no sign of life except when in the water, yet it is capable of continuing alive for many months though kept in a dry state.*

Perhaps, suggests scientist and expert on Romantic science Desmond King-Hele, Mary Shelley confused or mis-spelled Darwin's "vorticellae" as "vermicelli". Alternatively, she may have been thinking of a different note, in which Darwin does indeed discuss pasta spontaneously generating life:

> *... in paste composed of flour and water, which has been suffered to become acescent [sour], the animalcules called eels, Vibrio anguillula, are seen in great abundance; their motions are rapid and strong ... even the organic particles of dead animals may, when exposed to a due degree of warmth and moisture, regain some degree of vitality.*

Mary almost certainly read Darwin's work and she had ample opportunity to meet and talk with him at her father's London home, where he was a visitor. Yet a third possibility is that Mary, possibly via Darwin, may have read or heard one of Lucretius's own examples of spontaneous generation, for in *De Rerum Natura* the Roman writer discusses how rotting matter "brings forth worms", using the term *vermiculos* (worms/maggots). Could she simply have misheard or mis-spelled this word?

Alongside animated pasta, *The Temple of Nature* contains other suggestive lines relating to reanimation, such as Darwin's suggestion

Members of Birmingham's Lunar Society
meet for dinner; Erasmus Darwin was
one of the leading lights of the Society.

that regeneration from dead tissue might be a beneficial force
through some sort of global pleasure principle:

> *From the innumerable births of the larger insects, and the*
> *spontaneous productions of the microscopic ones, every part of*
> *organic matter from the recrements of dead vegetable or animal*
> *bodies, on or near the surface of the earth, becomes again presently*
> *re-animated; which by increasing the number and quantity of living*
> *organisms, though many of them exist but for a short time, adds to*
> *the sum total of terrestrial happiness.*

It is even possible to find in his lines clues to an explanation of
how the monster might have acquired a soul, for he suggests that
transmigration of the soul (a form of reincarnation expounded by the
Pythagoreans of ancient Greek Italy) is one facet of the "perpetual
transmigration of matter from one body to another, of all vegetables
and animals, during their lives, as well as after their deaths".

BOILED BROTH

Although spontaneous generation was still a current doctrine in Mary's time, it was on the defensive. The advent of scientific investigation, and particularly the increasing power of microscopes, had progressively depleted the range of creatures that were thought to arise from nowhere. English physician William Harvey, in his *De Generatione* of 1651, asserted that *ex ova omnia* ("all [life] comes from eggs") and identified invisible airborne "seeds" as the source of many cases of apparent spontaneous generation. In 1668, Italian

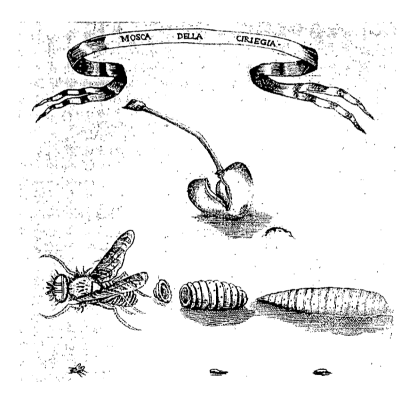

Illustration from Francesco Redi's seminal book,
Experiments on the Generation of Insects, *which helped to dispel several spontaneous generation myths.*

physician and pioneering parasitologist Francesco Redi showed that maggots come from the eggs laid by adult flies.

An almost fatal blow was struck to the doctrine of spontaneous generation in the eighteenth century after a dispute between the Comte de Buffon and his ally John Needham, a Catholic priest and accomplished microscopist, and the Italian priest and biologist Lazzaro Spallanzani. In 1748, at Buffon's urging, Needham had sought to buttress the evidence for spontaneous regeneration by showing that life would spontaneously arise in flasks of broth, even when they had apparently been sterilized by boiling and then kept sealed. Spallanzani suspected that Needham's safeguards against contamination might be inadequate, and in an elegant series of experiments, published in 1767, showed that the amount of growth of animalcules in a sealed flask of broth correlated to the length of time it was boiled: the longer the boiling, the less growth. A flask boiled for over half an hour displayed no signs of life, and when Needham objected that this was because such treatment had sterilized not only the seeds and spores that might have been present in the broth, but the very capacity of the medium to support life, Spallanzani proved him wrong by breaking the necks of the flasks and exposing the previously sterile medium to air, whereupon it soon grew cloudy with organisms.

Yet even this demonstration did not settle the matter, and the dispute over the true origin of organisms that appeared in supposedly sterile media rumbled on for over a hundred years, so that it was very much a live issue when Mary was writing *Frankenstein*. Spallanzani's work did, however, inspire the invention of food canning around 1810, when French chef Nicholas Appert applied the method to food sealed into bottles and boiled.

LOWER STATIONS

One of the final nails in the coffin of spontaneous regeneration came with the Pasteur–Pouchet controversy of the 1860s. Of a similar nature to the Needham–Spallanzani exchange a century before, the dispute centred on claims of spontaneous generation of germs in sterilized broth. Félix Archimède Pouchet (1800–72), Director of the

Natural History Museum in Rouen, claimed in his heterogenesis theory that life could spontaneously generate from organic matter, and that this was proven by his meticulously sterilized broth and flasks, which nonetheless displayed microbial growth.

When the French Academy of Science offered the Alhumbert Prize of 2,500 francs to whoever could shed "new light on the question of so-called spontaneous generation", microbiologist Louis Pasteur (1822–95) became interested. He identified the source of contamination in Pouchet's experiment as the mercury in the trough used for cooling, surmising that it trapped dust and spores from the atmosphere. Pouchet claimed, following Needham, that air was essential for heterogenesis, since it contained some vital principle. Accordingly, Pasteur designed an ingenious set of flasks

Pasteur at work in his laboratory; it was his work on atmospheric contamination of broth that helped to demolish the scientific credibility of spontaneous generation.

that allowed air to come in contact with the broth but trapped airborne particles, whence the broth remained clear for months. This proved, Pasteur reasoned, that the source of the inoculation of Pouchet's broth must have been either inanimate dust particles (for example, fabric fibres) or spores or eggs. "I prefer to think that life comes from life rather than from dust", he said, favouring the latter. Pasteur duly won the Alhumbert Prize, although typically the dispute with Pouchet rumbled on.

In the aftermath of this latest blow, the scope of spontaneous generation was limited to the lowliest forms of primordial slime, described in 1863 by Charles Darwin as "a fine hiding-place for obscurity of ideas", and by 1869 Joseph Lister was able to observe with disdain that the doctrine of spontaneous generation had been "chased successively to lower and lower stations in the world of organized beings".

ELECTRIC INSECTS: CROSSE'S SPONTANEOUSLY GENERATED MITES

Andrew Crosse (1784–1855) was an eccentric gentleman scientist whose dramatic and often noisy electrical experiments earned him nicknames such as "the thunder and lighting man" and "the Wizard of the Quantocks" (after the hills near his Somerset home). Sometimes cited as one of the inspirations for the character of Victor Frankenstein (*see* page 172), Crosse is primarily known today for his role in one of the more extraordinary cases of apparent spontaneous generation. Experimenting with the "electrical crystallization" of minerals, Crosse hooked up a lump of mineral iron oxide to a voltaic pile and dripped a solution of potassium silicate and hydrochloric acid on to it for several days, hoping to create crystals of natural glass.

PERFECT INSECTS

In this aim Crosse was unsuccessful, instead creating something much stranger. After fourteen days he "observed through a lens a few small whitish excrescences or nipples, projecting from about the middle of the electrified stone". Four days later these projections had "enlarged, and struck out seven or eight filaments", and by the twenty-sixth day, "These appearances assumed the form of a perfect insect, standing erect on a few bristles which formed its tail." Two days later the "little creatures moved their legs" and in a few more days "they detached themselves from the stone, and moved about at pleasure ... In the course of a few weeks, about a hundred of them made their appearance on the stone."

"I was not a little astonished," remarked Crosse. Assuming that his experiment had been contaminated to begin with, he started again, having meticulously cleaned his apparatus, but achieved the same result. The little bugs seemed to be of two types, one with six legs and one with eight, but as best as Crosse could tell they were of the genus *Acarus* (an arachnid mite related to spiders), "but of a species not hitherto observed". He offered little explanation for the

Some of Crosse's notes on the manifestation of his extraordinary Acari, with a sketch of the apparently electro-synthetic creature.

bizarre occurrence of the insects, able only to "suggest that they must originate in the electrified liquid by some process unknown to me".

WONDER-LOVERS

When his discovery was announced in a local newspaper, thence spreading around the country, the mites were dubbed *Acarus crossii*, and Crosse found himself plunged into controversy and infamy. Some attempts at replicating the generation of the mites were successful, but most were not. The mainstream scientific community was disdainful: Michael Faraday, the leading electrical scientist of his generation, who was erroneously described as having replicated the experiment, wrote to a friend: "With regard to Mr Crosse's insects etc. I do not think anybody believes in them here except perhaps himself and the mass of wonder-lovers." Other reactions were far more vicious: Crosse found himself subjected to vitriol, accusations of blasphemy and even death threats. "I have met with so much virulence and abuse ... in consequence of [these] experiments ... that I must state ... that I am [not] a 'self-imagined creator'", he protested, but to little avail. He suffered for many years as a result of the furore, becoming a virtual recluse thereafter.

HOUSE ITCH MITES

As for the mites, they are generally assumed to have resulted from contamination, with Crosse's efforts to eliminate this variable discounted as inadequate. One of the most curious features of the case was the presence of the unusual, crystalline filaments and hairs that characterized *Acarus crossii*, and which seemed to suggest that the little creatures shared both mineral and animal properties, and were indeed hybrids of the two kingdoms. In practice, the filaments may have been a clue as to the true identity of the bugs, since drawings of Crosse's mite resemble the acarid *Glycyphagus domesticus* – otherwise known as the furniture, storage or house itch mite – also characterized by long filamentous hairs. Thus, if Crosse's mites existed at all, they probably were indeed the result of contamination.

CHAPTER 4

OF MAN AND MONSTER

FRANKENSTEIN AND THE SCIENCE OF THE MIND

DRAMATIC CURES, MAGNETIC PERSONALITIES AND
EXCITING, RADICAL NEW THEORIES ABOUT MIND AND BRAIN,
CHARACTER AND CONSCIOUSNESS -- ALL THESE MARKED
THE BIRTH OF A NEW DISCIPLINE: PSYCHOLOGY. WHAT
DID IT MEAN FOR THE LINK BETWEEN THE ORGANIC AND
THE SPIRITUAL, THE WETWARE AND THE SOFTWARE? MARY
SHELLEY'S NOVEL WOULD EXPLORE THESE QUESTIONS AS
NEVER BEFORE OR SINCE.

THE NEW SCIENCE OF MAN: PHRENOLOGY AND THE DAWN OF PSYCHOLOGY

Victor Frankenstein does not simply make a mindless monster, an animated body – although the shadow of the automaton looms over the novel (*see* page 113). When he first conceives his plan to create life, he discusses not animals or organisms, but the "natures" he will midwife into existence. Accordingly he creates a *person*, a sentient being endowed with a powerful intellect and great sensibility. It was this, as much as the anatomical aspects of the monster, which animated, and has continued to animate, critical and popular response to the novel. Mary Shelley had created a psychology, and painted a compelling psychological portrait of it, yet psychology did not quite exist; like her monster, it was – at the very moment she was writing – being stitched together, brought ill-formed and monstrous into the world to excite extraordinary reactions. Shelley's creation reflected important scientific and ·social movements of her era, movements that had begun to create psychology as an entirely new and radically different branch of science: a science of man.

THE BRAIN IS THE ORGAN OF THE MIND

In 1815, the year before the composition of *Frankenstein* began, the word "phrenology" was coined by the British physician T. I. M. Forster; he derived it from Greek roots meaning "the study of the mind" – which were much the same roots as those that would later give rise to the word psychology. Phrenology was a practice, a pseudoscience and a social movement, predicated on core beliefs about the material nature of mind, the nature of mental faculties and the possibility of understanding such faculties by observing physical characteristics of the head.

Phrenology had developed from the work of the Viennese physician Franz Joseph Gall (1758–1828), a celebrated and expert anatomist who was the first to distinguish the functions of the "white" and the "grey" matter of the brain. Gall had anecdotally noted a link between bulging eyes and good memory, prompting a search for other correlations between mental faculties and physical features that led him to link directly the skull and the brain, in a theory he termed organology or simply "the physiology of the brain". Gall's anatomical prowess lent credibility to the system, but its propagation was primarily due to his student and acolyte Johann Kaspar Spurzheim (1776–1832), who became the great apostle for the new discipline.

The core tenet of phrenology was the doctrine that "the brain is the organ of the mind": a materialistic view locating the mind solely in the physical structure of the brain. Such views were increasingly commonplace in late-eighteenth-century science; Percy Shelley's doctor and mentor William Lawrence was a disciple of the German anthropologist Blumenbach (*see* page 69), who had employed his vast collection of skulls, known as "Dr B's Golgotha", to develop an elaborate craniology and insist upon the materialist, anti-dualist position that mind is purely a product of the physical structure of the brain (*see* page 109).

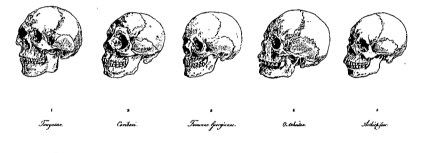

Illustration from one of Blumenbach's works on anthropology, which advanced a physicalist doctrine of mind alongside a pseudo-scientific racism.

THE DOCTRINE OF THE SKULL

More specifically, Spurzheim preached that distinct mental faculties, such as "destructiveness" and "benevolence", are localized in specific parts or "organs" of the brain, and that the size of such organs corresponded to the power and dominance of said faculty. Up to this point, there is not that much difference between the claims of phrenology and those of modern neuroscience, but infamously the phrenologists proclaimed the "doctrine of the skull": a belief in a direct link between the shape of the brain and the shape of the skull. According to Spurzheim: "the form of the internal surface of the skull is determined by the external form of the brain ... while the external surface of the skull agrees with its internal surface"; thus specific organs correlate directly to specific skull "protuberances", or bumps. Spurzheim further claimed that, through inspection and measurement of these bumps, it is possible to determine the strength of the mental faculties in an individual; in other words, the bumps on a person's head can be read to reveal their character, traits and abilities.

Gall had identified twenty-seven faculties, ranging from "larceny" and "religious sentiment" to "tenderness to progeny" and the "impulse to propagation". Under Spurzheim this increased to thirty-three, and inflated to as many as forty-three according to later phrenologists. Where Gall had favoured using his palms to "read" subjects' skulls, it became more common to use the fingers. Eventually, phrenologists would use scales to rate the strength of faculties, advise subjects which needed "cultivation" and which "restraint", and supply a manual of exercises to this end.

BUMPOLOGY AT THE PALACE

Spurzheim was an inspiring public speaker and effective popularizer, and he stressed the positive implications of phrenology, including the potential for self-improvement through training the faculties. Phrenology became a popular movement, even as it aroused opposition from the medical and scientific establishment. Among the most fervent proselytizers for Spurzheim's gospel was the Scottish moral philosopher George Combe, who termed phrenology the "science of man" and packaged it neatly for consumption by a growing constituency of self-educated middle and artisan classes with an appetite for a rational but accessible self-improvement system. By 1832, there were twenty-nine phrenological societies in Britain alone, along with journals such as Combe's own *Phrenological*

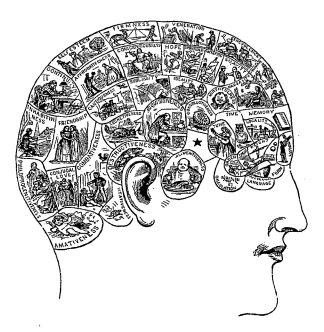

*A phrenological chart of the faculties of human
cognition; it was believed that correct stimulation (e.g.
by massage or phreno-mesmerism) might activate and
even enhance faculties.*

Journal, and in 1846 Combe was invited to read the bumps of the
royal children at Buckingham Palace.

Even by the time of *Frankenstein*'s composition, however, phrenology
and related practices – such as physiognomy (reading character from
the subject's face as well as skull) – were well known in Britain;
Mary Shelley herself had direct experience of these. As a three-
week-old infant she had been physiognomically diagnosed by a Mr
Nicholson, who reported on 18 September 1797 that Mary Godwin's
skull "possessed considerable memory and intelligence", while her
forehead, eyes and eyebrows displayed a "quick sensibility, irritable,
scarcely irascible" and her "too much employed" mouth showed "the
outlines of intelligence. She was displeased, and it denoted much
more of resigned vexation than either scorn or rage." In 1814 she
had learned about phrenology from her friend Henry Voisey, and
her journal makes it clear that she was familiar with Gall's system.

EYE OF THE BEHOLDER

Phrenology may not make an explicit appearance in *Frankenstein*, but its influence and assumptions run beneath the surface. The simple fact of the monster having a mind because Victor has given him a brain is testament to a strain of materialism or physicalism in the novel. His faculties may not be enumerated according to a strict phrenological prospectus, but they are manifest: his keen intelligence, his need for affection, his susceptibility to anger and violence. Are these innate faculties, determined by the nature of the brain that Victor has sewn into the cranium, or are they learned responses to his (absence of) nurture? Mary never discusses the provenance of the monster's brain, but it is telling that in later adaptations, such as the influential James Whale film of 1931 (adapted from a play), the creature becomes murderous because he has the brain of a criminal.

The craniological determinism of phrenology and physiognomy is perhaps more evident in the responses of other characters to the monster. Again and again his repellent appearance triggers assumptions

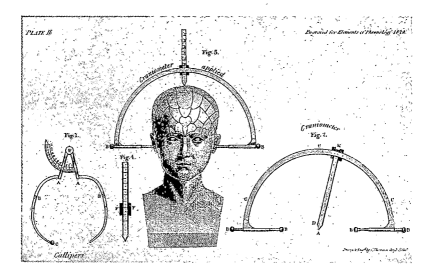

Illustration of a craniometer (device for measuring skull dimensions and bumps) from George Combe's 1824 Elements of Phrenology.

about his moral character and motivations; even the child William automatically attributes to him monstrous desires: "As soon as he beheld my form, he placed his hands before his eyes, and uttered a shrill scream ... 'Let me go,' he cried; 'monster! ugly wretch! you wish to eat me, and tear me to pieces – You are an ogre'". The only two people who do not reject the monster immediately on seeing him are the old man De Lacey, who is blind, and the polar explorer Walton, who, sympathetic because he knows the creature's biography, masters his instinctive loathing by shutting his eyes.

A MAGNETISM OF THE IMAGINATION: MESMERISM, HYPNOSIS AND CONTROL

Alongside phrenology, psychology's other late-eighteenth-century wellspring was mesmerism – a form of interpersonal influence attributed to animal magnetism and later conflated with hypnosis. The discourse of mesmerism stirred anxieties about manipulation, agency and control; the effects of mysterious energies and ethereal fluids; the threat that science might equip adepts with the power to usurp reason and autonomy. All these themes have their parallel in *Frankenstein*, reflecting the fascination and ambivalence of the Shelleys to this exciting but disturbing new technology of the psyche.

ANIMAL MAGNETISM
In the 1760s, the Viennese physician Franz Anton Mesmer (1734–1815) was intrigued by the apparently miraculous cures being effected by the public exorcisms performed by Catholic priest Johann Gassner, and sought to derive a naturalistic explanation using contemporary scientific theories. Initially he turned to Newtonian physics, and won his doctorate in 1766 with a thesis that attempted to apply gravity to human physiology, using a theory of "animal gravity". After Mesmer encountered the therapeutic use of magnets, animal gravity became animal magnetism, a force akin to but different from

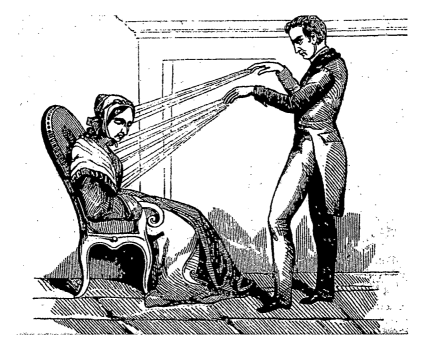

The classic depiction of a mesmerist working his dark arts on an entranced subject.

normal magnetism, conceived as an etheric fluid flowing through and between living things, imbalances in which cause mental and physical illness. Manipulation of this animal magnetic fluid – initially with magnetized rods and other devices, and later simply with his own hands – could redress imbalances and produce remarkable cures, both physical and psychological, often by inducing and then resolving a "crisis": a moment of physical and emotional climax.

Hostility from the medical establishment forced Mesmer out of Vienna, but his move to Paris in 1778 brought still greater success and celebrity. To enable treatment of multiple patients he devised his *baquet* – an oaken barrel or tub filled with therapeutic items such as bottles of "magnetized" water and aromatic herbs, from which projected iron rods and ropes. This device powered a human circuit of animal magnetism dubbed the "magnetic circle", while Mesmer manipulated the flow of animal magnetism by passing his hands over patients and staring intently into their eyes. His clientele was mainly female and there was an obvious element of sexual frisson, which in turn led to accusations of impropriety.

Imitators began to spread the gospel of mesmerism, and an alarmed medical/scientific establishment retaliated by pushing a Royal Commission headed by American scientist Benjamin Franklin (*see* page 37), and featuring both Antoine Lavoisier and the man whose eponymous device would later cut off his head, Joseph-Ignace Guillotin. In a classic example of the application of the scientific method to medical trials, the commission debunked Mesmer's claims and showed that the effects he produced, though real, were due to suggestibility and expectation; what would now be called the placebo effect. Discredited, Mesmer left Paris and faded into obscurity.

DARK POWERS

Nevertheless, the practice of mesmerism endured, with successive waves of popularity ensuring that it spread to Britain and the United States, where it became especially popular and widespread in the early Victorian period. Animal magnetism may not have gained the imprimatur of science, but it remained an influential force firmly embedded in the popular imagination, paralleling the public reception and conception of electricity. Both forces were touted as therapeutic agents, and both reached large audiences through lecture tours that combined theatrical entertainment with interactive demonstration.

As with the use of electricity – specifically galvanism – mesmerism was also shadowed by darker reflections, engendering anxieties around the fear of violation and control. Both appealed to explanations involving ethereal fluids and animal energies, which could be manipulated by powerful and charismatic individuals to threaten autonomy and even identity. More widely, mesmerism fed into a growing anxiety about the power of science and technology: their potential for evil and unintended consequences in the hands of hubristic scientists – a trope with obvious significance for the composition of *Frankenstein*.

SOMNAMBULISM

Mesmerism would eventually morph into hypnotism, by way of what was initially termed "magnetic sleep". This was a trance-like condition that sometimes affected Mesmer's subjects, a phenomenon to which he paid little attention but which fascinated his most important follower, the Marquis de Puységur (1751–1825). Noting the similarities of this state of extreme suggestibility and passivity to waking dreams, Puységur later termed it "somnambulism": an apparently distinct state of consciousness in which a person acts like a sleepwalker.

Experiments convinced Puységur that somnambulists had psychic powers, and also revealed that subjects had no memory of their experiences while in a trance. Focus on the trance state thus shifted mesmerism from a primarily somatic phenomenon to a psychological one, and this was later taken up by the Scottish physician James Braid (1795–1860), who coined the term "hypnosis", from the Greek for "sleep". Of particular interest in the context of *Frankenstein* is that Percy Shelley was a sleepwalker; an altered state of consciousness that, like mesmerism, raised questions of volition, autonomy, identity and automatons. It is also curious to note that the doctoral thesis of Polidori, the physician who was among the party at the Villa Diodati and who must, to some extent, have contributed to the genesis of *Frankenstein*, concerned somnambulism.

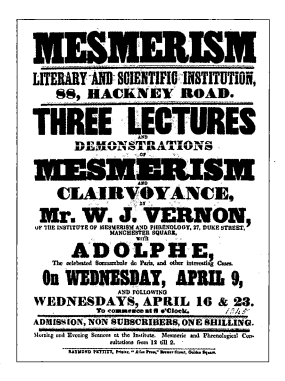

Advertisement for a show by a mesmerist and phrenologist promising
demonstrations of altered states and psychic powers.

A TREMENDOUS SHIVERING OF THE NERVES

Mary had first-hand experience of the anxiety around mesmerism, through the person of her lover, Percy Shelley. Fascinated with ethereal fluids such as animal electricity and animal magnetism, Percy was also disturbingly vulnerable to them. Poets were expected to display heightened "sympathies", both with nature and with other people, but this could tip over into dangerous territory. His cousin Tom Medwin memorably described Percy as excruciatingly "sensitive to external impressions", and as "so magnetic" that he could become overpowered by emotions and feelings emanating from others. "When anything particularly interested him, [Shelley] felt a tremendous shivering of the nerves pass over him, an electric shock, a magnetism of the imagination." The Shelleys' joint anxieties about external influence and loss of control would manifest themselves in strange dreams and terrifying visions of doppelgangers (*see* page 112).

The year before Mary came to write her novel, the German Romantic E. T. A. Hoffman published his story "The Magnetiser", an eerie and disturbing tale about a beautiful young woman who submits to a "magnetic cure" at the hands of a mesmerist, only to find that he has enslaved her to his will, "directing her innermost thoughts through magnetic means", with tragic ends. The diabolical Magnetizer is even able to cause her to drop down dead: a kind of inverse Frankenstein, able to take away life through his mastery of vital forces. It is unclear whether Mary Shelley read Hoffmann; her German was not strong enough to read the original, but accounts of the party at the Villa Diodati indicate that the group were reading and discussing other German tales, so it is not beyond the bounds of possibility.

PAY IT FORWARD

Frankenstein would eventually feed back into the discourse on mesmerism, its imagery and tropes co-opted by mid-nineteenth-century mesmerists – particularly via its stage adaptations. According to Alison Winter, in her book *Mesmerized: Powers of Mind in Victorian Britain*, "Audiences often recalled [*Frankenstein*] when reflecting on the issues mesmerism raised." She details a broadsheet of January 1846 advertising the lecture/entertainment offered to the people of Huddersfield by a Mr W. Richardson, illustrated with a dramatic image of lightning bolts playing around the pointed towers of a Gothic castle, and promising an electro-magnetic engine that would give "the appearance of LIFE TO A DEAD BODY".

PHRENO-MESMERISM

Perhaps inevitably, the twin movements of proto-psychology would become aggregated into a single phenomenon: phreno-mesmerism. In around 1843, practitioners began to combine the two: phrenology provided a guide to the identity and location of mental faculties, while mesmerism gave the power to "supercharge" them. Placed in a mesmeric trance, individuals could have their phrenological faculties activated and enhanced. An Edinburgh clergyman prompted his entranced daughter to pray by "exciting" her "phrenological organ" of veneration, and to tear her dress in a frenzy when he excited her destructiveness. Similarly, an artisan in Nottingham who had mesmerized his maid prompted her to make a confession by touching her organ of conscientiousness.

THE EDUCATION OF A MONSTER: THEORIES OF COGNITIVE AND PERCEPTUAL DEVELOPMENT

The modern icon of Frankenstein's monster as a lumbering mute derives from the motion picture adaptations, placing it radically at odds with the monster of the novel itself, who is articulate, sensitive and highly intelligent. The central portion of the book gives the creature's own, detailed account of his intellectual and moral development, an account that reflects the philosophical debates of the era and Mary Shelley's interpretation of contemporary developmental psychology. Her monster provides a kind of imaginary laboratory for working through contesting theories of cognition; indeed, his status as a thought experiment – or, more properly, a whole series of interlinked thought experiments – is one of the reasons that the novel is often lauded as the first true work of science fiction.

PAPER AND MARBLE

In Mary's time thinking about cognition and its development – in philosophical terms, epistemology, the study of the origin and nature of knowledge – was dominated by the debate between rationalism and empiricism. Rationalism, championed by Descartes, argues that the fundamental building blocks of knowledge are innate to the mind, dependent on reason alone and not on the external world. Empiricism, championed by the English philosopher John Locke (1632–1704) in his 1690 *Essay Concerning Human Understanding*, argues that knowledge comes exclusively from experience, and that the mind comes to know things through reflecting on sensory input.

In particular, Locke is associated with the description of the initial state of the mind as a *tabula rasa* (Latin for blank slate). This is a phrase deriving from Aristotle, but it does not actually occur in Locke's *Essay*, appearing initially in a French translation ten years later. What Locke actually said bears a distinct resemblance to the language Mary Shelley would later use to describe the monster's early cognitions:

> *How does [man] come by [ideas]? Let us suppose the mind to be as we say, white paper, void of all characters, without any ideas. Whence has it all the materials, of reason and knowledge? From experience ..., our senses ... do convey unto the mind several distinct perceptions of things according to these various ways wherein those objects do affect them ... yellow, white, heat, cold ... [this] I call sensation. Secondly ... reflection, the notice which the mind takes of its own operations.*

Compare this to an example taken from the creature's account of his early days: "My sensations had, by this time, become distinct, and my mind received every day additional ideas. My eyes became accustomed to the light, and to perceive objects in their right forms; I distinguished the insect from the herb, and, by degrees, one herb from another." He begins with sensation, and moves on to reflection (evidenced by his developing ability to categorize his perceptions). So, Mary's account of the monster's cognitive development seems clearly to favour an empiricist reading.

More specifically, she seems to have closely followed the theories of the English empirical materialist David Hartley (1705–57) and the French empiricist philosopher Etienne Bonnot de Mably de Condillac (1715–80). Hartley attempted to apply Newtonian-style physics to Locke's empiricism and so develop a biological basis for the *tabula rasa* theory, and Mary's account of the monster's cognitive evolution mirrors closely the programme set out in Hartley's 1749 *Observations on Man, His Frame, His Duty and His Expectations*. Condillac similarly sought to develop Locke's empiricism in his 1754 *Traité des sensations*, in which he likens the individual learning to understand sensory input to a marble statue, equipped with sensory organs. Empiricists believed that sensory input made physical impressions on the nervous tissue, a mechanism that seems to operate on the monster.

NATURAL LANGUAGE

The novel supplies its own answer to a celebrated thought-experiment related to epistemological debates: what language will children speak if raised without language? According to legend, several attempts were made to answer this question in the real world. For instance, in 1493, according to the historian Robert Lyndsay of Pitscottie, King James IV of Scotland had two infants and a deaf mute nurse transported to the isolated island of Inchkeith, in the middle of the Firth of Forth. Lyndsay wrote, "Some say they spoke good Hebrew; for my part I know not, but from report", although as the novelist Sir Walter Scott later observed, "It is more likely they would scream like their dumb nurse, or bleat like the goats and sheep on the island." Mary Shelley seems to have agreed with Scott's diagnosis, for the monster, left to his own devices, can produce only "uncouth and inarticulate sounds" that frighten him into silence. Only when he can overhear humans talking is he able to learn language, albeit with amazing rapidity.

THE COMMON SENSORIUM

*The philosopher Jean-Jacques Rousseau, whose
thinking about human nature and cognitive
development cast a long shadow over* Frankenstein.

Another philosopher whose influence looms large
in *Frankenstein* is the Swiss-born Francophone Jean-
Jacques Rousseau (1712–78). Among his best-known
theories, despite the fact that he never used the term,
is that of the "noble savage", elements of which clearly
inform the character of the monster. A lesser-known
but intriguing line of influence comes from Rousseau's
theory of cognitive development, in particular a
passage in which he appears to anticipate the monster:

[if] a child had at its birth the stature and strength of a man, that he had entered life full grown like Pallas from the brain of Jupiter ... all his sensations would be united in one place, they would exist only in the common "sensorium".

In modern terms this sensorium or "unification of the senses" is known as synaesthesia, a phenomenon in which different sensory modalities intertwine to produce a sensory synthesis: musical tones might be perceived as having colours, shapes might have distinct tastes or smells, and so on. Textual evidence shows that Mary Shelley may have taken direct inspiration from Rousseau, for the monster seems to be describing synaesthesia in his account of his initial sensory development:

A strange multiplicity of sensations seized me, and I saw, felt, heard, and smelt, at the same time; and it was, indeed, a long time before I learned to distinguish between the operations of my various senses.

In a 2007 paper for the Society of Neuroscience in San Diego, researchers A. O. Holcombe, H. J. Over and E. L. Altschuler note this link and draw a further parallel with a modern theory about sensory/cognitive development in infants, which suggests "that we all begin life as synaesthetes; developmental pruning of neural connections eventually yielding more segregated senses". According to this theory, cognitive development is not, or not only, the result of growth of neural connections, but actually of pruning away an initial oversupply of connections, like cutting back an overgrown bush to produce a distinct shape. The monster's own cognitive development certainly seems to fit with this model, as he learns to separate the initial inchoate mass of perceptions into discrete categories and modalities.

EXAMINE THE MIND

Erasmus Darwin and William Lawrence espoused related theories, and were both major influences on Mary Shelley (*see* pages 76 and 170). Darwin advanced a materialistic (albeit slightly mystical) view of the origins of emotion and sensation, particularly pleasure and pain, describing them as functions of the expansion and contraction of nerve and muscle fibres: "the actions of the organs of sense" on "the fibres which perform locomotion". Lawrence, in a lecture of 1817, asked his fellow members of the Royal College of Surgeons to "examine the 'mind'!":

> *Where is the "mind" of the foetus? Where is that of the child just born?*
> *Do we not see it actually built up before our eyes by the actions of the*
> *five external senses, and of the gradually developed internal faculties?*
> *Do we not trace it advancing by a slow progress from infancy and*
> *childhood to the perfect expansion of its faculties in the adult ..."*

This sounds very much like a-description of Chapter 11 of *Frankenstein* – but with one obvious difference. Where Lawrence traces the advancement of the mind by slow progress from infancy to adulthood, Mary's monster completes his cognitive development with astonishing rapidity. This may be simply a literary device to allow the plot to advance, but it makes the monster the first in a long line of similar characters in science fiction: newly created beings who are able to progress with astonishing rapidity from childlike naivety to advanced intelligence and encyclopaedic knowledge – see, for instance, the Luc Besson film *The Fifth Element*.

THE EMPEROR OF CHINA: IDENTITY, MIND AND BODY

Analysis seeking the source of the novel's enduring power to fascinate and discomfort often points to the question of the creature's soul. Victor Frankenstein has created a material body and even equipped it with a functioning brain and an able mind. But the creature is more than just an animal or a machine; he has personality and character, dreams and desires, philosophy and metaphysics. If there were a Turing test for soul, he would surely pass it. Yet he was created not by God or nature, but by a flawed man; forged not in a supernatural creation scene or the womb of a mother, but cobbled together from rotting body parts. Could he really have a soul, and if so, what does this mean for other souls? Can souls be artificially created and loaded into the wetware of revivified tissue, like software written by a programmer and loaded onto computer hardware? Such an analogy was not available to Mary Shelley or her contemporaries, but these concerns certainly were, and her novel explores like none before issues of dualism, the mind-body problem and the philosophy of identity.

DESCARTES' DAUGHTER

Descartes' mechanistic view of physiology (*see* page 40) left him with a problem when it came to human consciousness. If physiology and all the operations of the body are purely material, physical phenomena, then this must also apply to the brain, the organ of the mind. Did mechanism thus lead to a materialist view of the mind, effectively denying the existence of the immaterial, including necessarily the immaterial soul? This would be atheism, a charge that Descartes was quick to reject. He sought to rescue the immaterial soul by positing a clear divide between it and the material brain, between mind and body; they are made from two different types of stuff. Hence this position is known as dualism. Descartes suggested that the bridge between the two worlds – for it was impossible to deny that mind interacted with body – lay in the pineal gland, which functioned as some sort of inter-dimensional gateway.

Yet suspicion about the materialist implications of Descartes' mechanism lingered, to the extent that there arose a macabre legend about him after he died. It was said that he had travelled to Sweden, boarding the ship with his daughter Francine, but that once under way the crew did not see the girl. Becoming alarmed they opened Descartes' luggage and found an eerily life-like mechanical doll, which they threw overboard. In fact Descartes' daughter Francine had died in 1635, at the age of five, but there was much in his work to lend credence to a fascination with automata (*see* page 113).

LEIBNIZ'S MILL

Descartes' dualist rejection of the material nature of consciousness received support from the German polymath and philosopher Gottfried Wilhelm Leibniz (1646–1716). He denied that an understanding of material "mechanics" (including the workings of the body and brain) could fully explain the mental world, that is, thinking and consciousness, which he sometimes referred to with the term "perception". "Perception, and what depends upon it," he insisted, "is inexplicable in terms of mechanical reasons, that is through shapes, size and motions", illustrating his contention with a thought experiment:

> *If we imagine that there is a machine whose structure makes it think, sense, and have perceptions, we could conceive it enlarged, keeping the same proportions, so that we could enter into it, as one enters a mill ... when inspecting its interior, we will find only parts that push one another, and we will never find anything to explain a perception.*

Leibniz's enlargement of his machine, to enable us to cope with its complexity, seems to find a parallel with Victor Frankenstein's decision "to make the being of a gigantic stature" owing to "the minuteness of the parts". If the monster's brain were large enough to "enter into" and "inspect", would we see anything to "explain a perception"? Leibniz would contend not, because his argument was not simply that a machine cannot think, but that the brain itself cannot be sufficient to explain consciousness.

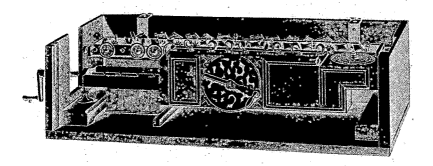

*Leibniz's "step reckoner", an early mechanical calculator;
despite espousing a dualist position, Leibniz also
pioneered the philosophy of machine intelligence.*

MACHINE MAN

Others rejected Descartes' dualist dodge and embraced the full
implications of his materialism, including a philosopher who greatly
influenced Percy Shelley. This was Julien Offray de La Mettrie (1709–
51), a French physician who proposed an extreme mechanistic view
of body and mind. He dismissed the soul as "an empty word to which
no idea corresponds" and, with a strikingly prescient observation,
used the effects of brain lesions to demonstrate the essentially organic
origin of cognition and directly undermine Descartes:

> since diseases of the brain, according to the place they attack,
> destroy sometimes one sense and sometimes another, are those who
> place the seat of the soul in one of the pairs of optic lobes any more
> wrong than those who would like to limit it to the oval centre, the
> corpus callosum or even the pineal gland?

In his most radically mechanistic work, *L'Homme-Machine* (Man a
Machine, 1747), he described the human body as "a machine that winds
itself up" and men as "perpendicularly crawling machines". He did not deny
that humans have minds, but ascribed the origin of mind to the material

substrate of the body, which in humans achieved a level of such complexity and sophistication that consciousness emerged. Here, La Mettrie once again proves remarkably prescient: an early prophet of the modern doctrine of consciousness as an emergent property of complex systems.

According to the Irish writer and literary historian Seamus Deane, Percy Shelley was more influenced by French Enlightenment materialism, and La Mettrie in particular, than "any other English writer of the period 1789–1832". Mary must have imbibed, through Percy, some flavour of this mechanistic philosophy, and certainly on the face of it her monster stands as a clear statement of materialism: an assemblage of body parts that, when brought to life by natural means, is automatically equipped with a fully functioning psyche.

DOPPELGANGERS

The terminology of P-zombies (see opposite page) was not available to Mary and Percy Shelley, but they were animated by similar concerns; Percy in particular suffered from dread of the doppelganger, a double that threatened to subsume his identity and perhaps usurp his consciousness. Would such a doppelganger be a P-zombie, or would it leave him as a P-zombie? These fears bled over into Mary's novel, in which the monster becomes Victor's dark doppelganger, a being that he created but which has first developed autonomy and now threatens to usurp his own. This link is made explicit in the novel, as when Victor, writhing in horror as he considers what he has done, mimics the vital stirrings of the creature: "my teeth chattered, and every limb became convulsed"; or when he is moved to quote Coleridge's lines about being haunted by a doppelganger:

Like one who, on a lonely road,
Doth walk in fear and dread,
And, having once turned round, walks on,
And turns no more his head;
Because he knows a frightful fiend,
Doth close behind him tread.

THE CREATURE AS A ZOMBIE

Although Descartes allowed humans a dual nature, he denied this to "brute" beasts. Animals, he argued, are indeed nothing more than meat machines, automatons without consciousness. Considering the possibility of a human automaton, Descartes dismissed it on the basis that such a creature could not be creative with language or behaviour. He asserted that if a human lost its consciousness, its body might continue to work, able to walk and even sing in a mindless fashion, but it would no longer be a person. Modern philosophy has been much less certain, however, that it would be so easy to distinguish a conscious person from what Descartes might have called a human automaton, but which in contemporary parlance is known as a philosophical zombie or "P-zombie".

A P-zombie is not like the living dead of horror movie fame. It is a hypothetical being who walks and talks exactly like a normal human, demonstrating the same behaviours, reactions, emotions and speech as a "real" person, even down to the level of displaying patterns of neural activity indistinguishable from a non-zombie. The only difference is that the P-zombie lacks conscious experience,

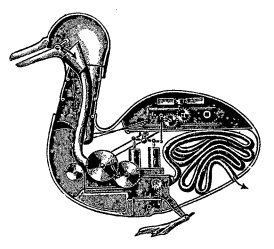

Cutaway schematic of one of the mechanical ducks of eighteenth-century inventor Jacques de Vaucanson; his remarkable animatronics seemed to blur the line between organic and artificial life.

also known to philosophers as intentionality or qualia. A conscious being does not only behave as if feeling happiness; he or she also has a conscious, subjective experience of happiness. There is something that it is like for them to feel happiness, whereas for a zombie there is nothing that "it is like" for any experience. An animatronic doll programmed to smile might look exactly as if it is happy, but in reality it feels nothing.

Philosophers posit P-zombies as a tool for critiquing physicalism – the notion that mind can be explained purely in physical terms. P-zombies are physically identical to humans, yet this is not sufficient to produce consciousness. Some philosophers argue that the fact that we can even imagine such a thing as a P-zombie is proof that physicalism is wrong.

The monster is a purely physical creation. If physical facts alone are not sufficient to engender consciousness, then the monster should be a P-zombie. But he fairly clearly is not – several chapters are devoted to detailing his subjective experiences and he has plenty of intentions – which suggests that Mary Shelley was espousing a physicalist philosophy of mind.

THE INDIVIDUAL ANNIHILATED

Accepting that the monster does have a soul, at least in the sense of being a person with conscious experience, subjectivity and intentionality, leaves open the question of where this soul comes from. The creature has a personality and character, but whose? From where does his identity derive? The most obvious possibility is that it came with his brain: a fully formed, adult brain, taken from the body of a person who also had an identity. Is the monster, then, basically the same person as whoever "donated" his brain to Victor's home science project?

Questions such as these highlight how the novel challenges conceptions of identity and continuity of identity, another classic theme in the philosophy of mind. One attempt to answer them came from Leibniz, with a thought-experiment that seems highly applicable:

> Let us suppose that some individual suddenly became the Emperor of China, but only on condition that he forgot what he had been,

> *as if he had just been reborn: does that not come to the same in*
> *practice, or in the effects that could be registered, as if he had to be*
> *annihilated and an Emperor of China created at the same instant*
> *at the same place?*

Substitute "Frankenstein's monster" for the "Emperor of China" and Leibniz's scenario is a perfect fit. Common sense, Leibniz proposes, insists that we consider the monster to be an entirely new identity, with no debt to the former owner of the brain.

THE SECOND SHIP

An alternative approach is suggested by one of the oldest known thought-experiments, the paradox of the Ship of Theseus. This was an ancient wooden ship preserved in the city of Athens, and said to have been the same vessel that bore Theseus home from his scrape with the Minotaur. When a plank rotted away it was replaced with fresh timber, causing philosophers to ponder whether the ship remained the same. If every timber were replaced so that nothing original remained, would it still be the same ship?

The seventeenth-century English philosopher Thomas Hobbes (1588–1679) suggested a spin on the Ship of Theseus that is relevant to *Frankenstein*. He imagined a second ship constructed from the discarded planks of the first; this ship would be, in one sense, new, but in another sense more authentic than the first ship, since it was made from the original materials. Victor constructs his monster from discarded body parts; if the monster were reconstructed from parts of one individual, would he be that individual? Given that the monster is actually constructed from multiple different individuals, does he partake in any of their natures? Mary appears to argue that human tissue acts as a sort of existential palimpsest: once dead, the tissue loses any former identity and can be recycled into an entirely new identity, but the open question remains: from where does this identity, this character, this soul, arise?

DESCARTES' EVIL GENIUS

Perhaps the most famous thought-experiment to come from seventeenth-century philosophy is Descartes' "evil genius". In an era of growing scepticism about ancient and established authority, Descartes wondered how it is possible to know that anything is true: "Suppose some evil genius not less powerful than deceitful, has employed his whole energies in deceiving me; I shall consider that ... all ... external things are but illusions and dreams of which this genius has availed himself to lay traps for my credulity." Recast in a modern guise this is also known as the "brain in a vat" hypothesis: how do you know that you are not simply a disembodied brain floating in a vat, hooked up to a supercomputer that feeds you an utterly convincing virtual reality, as in the film *The Matrix*? Descartes' response is summed up in the famous line, "*cogito ergo sum*" – "*I think, therefore I am*" – although he did not use this actual phrase. What this signifies is that while you may doubt everything, you can at least be sure that there is an essential self that is doing the doubting.

The monster may not be floating in a vat, but there are clear parallels between the novel and Descartes' scenario. In this reading Victor is the evil genius, the powerful entity who has chosen to reactivate an inanimate brain and thus to control the phenomena of consciousness. But the monster, like Descartes, asserts the irreducible individuality of his consciousness.

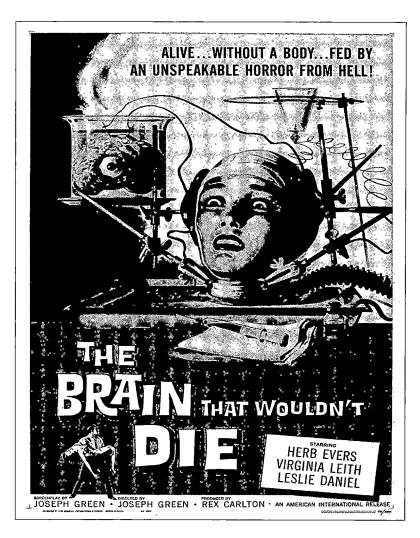

Before The Matrix *there was ...* The Brain That
Wouldn't Die, *a 1962 horror film riffing on the
brain-in-a-vat thought experiment.*

ANDROID INTELLIGENCE

Frankenstein's monster is often cited as the progenitor of a prolific fictional tribe: the androids. Though made of organic parts rather than metal and silicon, the monster is a material construction in the form of a man. Do his claims to an authentic human consciousness then face the same objections as the artificial intelligence of an android? The philosophy of artificial or machine intelligence is a rich and diverse field, with many implications for the dualism versus materialism debate at the heart of Mary Shelley's novel. The claim that a machine could, even theoretically, have the same kind of intelligence as a human has spawned multiple critiques. One prominent challenge is the Chinese Room argument of the American philosopher John Searle (b.1932).

Searle likened the intelligence of a machine to a closed room, inside which is seated a man who does not speak, read or understand Chinese. His only channel of communication with the outside world is a slot through which messages are posted, written in Chinese characters. The man follows instructions set out in a huge book, which direct him to process the characters in such a way as to generate a new set of characters, which he copies and posts back through the slot. This leads to a paradox: to the Chinese person outside the room, the man in the room appears to be able to read and write Chinese, whereas in fact the symbols are meaningless to him and he simply follows instructions. A machine intelligence would be the same as the man in the Chinese Room, able to process symbols to generate responses meaningful to interlocutors, but with no subjective comprehension of their meaning.

An organic human brain, Searle contends, is different from a machine, for any machine construction "necessarily leaves out the biologically specific powers of the brain to cause cognitive processes". He stresses that mental states are real biological phenomena, grounded in embodiment and interaction with the physical world, in a similar fashion to digestion or photosynthesis. This special pleading for the importance of organic "wetware" has been called carbon or protoplasm chauvinism. What happens with a construction like Frankenstein's monster, with his embodied organic brain? Would Searle allow the Chinese Room paradox to be avoided if the machine is made of meat, as is the case with the monster?

CHAPTER 5

ANATOMY OF HORROR

DISSECTION, MURDER AND RESURRECTION IN THE ROMANTIC ERA

THEN AND NOW, MUCH OF THE IMPACT OF THE NOVEL STEMMED FROM THE BODY HORROR, BOTH EXPLICIT AND IMPLICIT, OF ITS NARRATIVE. VICTOR FRANKENSTEIN "DABBLES AMONG THE UNHALLOWED DAMPS OF THE GRAVE" AND "COLLECTS BONES FROM CHARNEL HOUSES" TO GATHER BODY PARTS FOR A MONSTROUS CREATION THAT HE STITCHES TOGETHER. ITS FINAL FORM IS TOO REPULSIVE TO LOOK UPON WITHOUT RECOILING. PHYSICALLY AND METAPHYSICALLY IT INVOKES DISGUST AND TERROR, TRANSGRESSING THE BOUNDARIES OF LIFE AND DEATH, VITAL AND CORRUPT. WHAT ARE THE ORIGINS OF THIS VISCERAL HORROR? THEY CAN BE TRACED TO THE MACABRE FASHIONS AND FASCINATIONS OF AN AGE IN WHICH NO SEPULCHRE WAS FREE FROM THE SHADOW OF THE RESURRECTION MEN, ANATOMISTS PERPETRATED HIDEOUS OUTRAGES UPON DISMEMBERED CORPSES, AND THE LINE BETWEEN LIFE AND DEATH THREATENED UTTERLY TO DISSOLVE.

THOSE FLAYING RASCALS: BODY-SNATCHERS AND DISSECTIONS

Recalling his macabre research, Victor Frankenstein employs to the fullest extent the rhetoric of the Gothic. "Who shall conceive the horrors of my secret toil, as I dabbled among the unhallowed damps of the grave, or tortured the living animal to animate the lifeless clay?" he asks, remembering how he "collected bones from charnel houses; and disturbed, with profane fingers, the tremendous secrets of the human frame ... The dissecting room and the slaughterhouse furnished many of my materials; and often did my human nature turn with loathing from my occupation." From the most base, Victor seeks to bring forth the sublime, but his "secret toil" would have held grim resonance for readers in the early nineteenth century. Even as Mary Shelley gestated her novel, the resurrection men and the body-snatchers plied their "loathsome occupation". How do the fictional exploits of Victor Frankenstein compare with those of the real thing?

COMPANIONS WITH THE DEAD

Tim Marshall, author of *Murdering to Dissect: Grave-robbing, Frankenstein and the Anatomy Literature*, describes *Frankenstein* as "the classic story of the bodysnatching era", explaining that "Victor Frankenstein combines the roles of resurrectionist and anatomist". What were these roles, and how does Shelley's novel reflect contemporary understanding of their work?

In 1829, an eminent surgeon, and former Professor of Anatomy at the Royal College of Surgeons, told students at St Bartholomew's Hospital, "[t]here is but one way to obtain knowledge [of human anatomy] ... we must be companions with the dead". Some eighteen years earlier Percy Shelley had attended the very same lecture course, having become enamoured of anatomy and considering a career in medicine; it is possibly no coincidence that in the novel Victor Frankenstein makes similar claims:

> *To examine the causes of life, we must first have recourse to death*
> *... I became acquainted with the science of anatomy: but this was*
> *not sufficient; I must also observe the natural decay and corruption*
> *of the human body.*

Abernethy and Victor were espousing an ancient but contested tradition. Ancient Greek and Roman religions proscribed the "desecration" of the bodies of the dead, but philosophers and physicians transgressed such rules since at least the sixth century BCE, when the Greek philosopher Alcmaeon was the first recorded to have performed medical dissections. Herophilus of Chalcedon (a follower of Hippocrates) founded the first school of anatomy at the Museum of Alexandria (the pre-eminent academic institution in the Hellenic world), in 275 BCE, exhorting his students to overcome their fear of dissecting the dead. In 30 CE, the Roman physician Celsus, in a review of Greek medical texts, insisted that "to open the bodies of the dead is necessary for learners". Galen, a second-century CE Graeco-Roman physician who became the pre-eminent medical authority of the Middle Ages, was said to have performed two secret human dissections, although his anatomical legacy would become highly contested.

In the medieval period, the proscription on human dissection strengthened, as it clashed with mainstream interpretations of the doctrine on resurrection (*see* box on page 128), and in 1163 the Council of Tours led to a formal prohibition against human dissections. However, by 1315 the Vatican had relaxed its rules enough to sanction the first recorded public demonstration of human anatomy, by the Italian physician and surgeon Mondino de Luzzi, employing the corpse of an executed female criminal.

During the Renaissance, artistic and natural philosophical imperatives combined to increase interest in human anatomy. Leonardo da Vinci (1452–1519) stressed the importance of studying anatomy and made important contributions to comparative anatomy, in which parallels were drawn between human and animal anatomy. The greatest influence on anatomy, however, would be the sixteenth-century Flemish anatomist and physician widely considered the father of anatomy, Andreas Vesalius (1514–64), who used his personal experience of human dissection to overturn inaccurate Classical authorities. He showed, for instance, that much of Galen's work was wrong, owing more to guesswork and animal anatomy than accurate human anatomy.

Vesalius' monumental 1543 work *De corporis humani fabrica* provided the first widely available accurate record of human anatomy – yet he still suffered from religious taboos and was sentenced to death by the Inquisition for his research. But the Reformation limited the authority of the Catholic Church and in Protestant countries physicians were allowed greater freedoms. The Royal College of Physicians in London, for example, was authorised to perform human dissections in 1565.

Illustration from a 1541 edition of Galen's works, showing the ancient physician anatomizing a pig; his claims about human anatomy were shown to be suspect.

EXCESSIVE APPLAUSE

Human dissection became – and remains to this day – an essential part of medical education. But anatomy demonstrations were of more than merely academic appeal; they became a form of popular theatre. A particularly prominent venue was the anatomy theatre of the Royal College of Surgeons in London. Astley Cooper, a leading surgeon who performed the public dissections there between 1793 and 1796, recalled that the theatre was "constantly crowded, and the applause excessive".

Audiences could be treated to macabre spectacles, as with Aldini and Ure's galvanic torment of the corpse (*see* pages 54 to 59), William Clift's assault on the eyeballs (*see* page 137) or a grisly fascination with post-mortem cardiac activity. Hearts were observed in situ, prodded and shocked and even cut out and placed in a dish to be looked at. When the heart finally stopped pulsing, it might be poked with a scalpel to see if it would start again.

By the late seventeenth century, when this engraving was produced, human anatomy was a legitimate object of study; the heart would prove of particular interest to anatomists at the Royal College of Surgeons.

AMONGST THE OTAMIES

With dissection now legal, and increasingly central to medical schooling, demand for human corpses became exigent. In England the 1751 Murder Act (or, to give its full name, "An Act for better preventing the horrid Crime of Murder") legislated that the bodies of those hanged for murder be given over to anatomists for dissection. However, not only did the supply fail to satisfy the demand, the law had unintended consequences. Dissection was viewed as worse punishment than execution, and the anatomists and their henchmen, waiting by the gallows to carry the corpses direct to the operating theatre for maximum freshness, became figures of hatred and fear. There were increasing outbreaks of violence after hangings, as the mob sought to prevent the bodies being taken. Eighteenth-century English novelist Samuel Richardson recorded the aftermath of an execution:

> As soon as the poor creatures were half-dead ... the populace [fell] to hauling and pulling the carcasses with so much earnestness, as to occasion several warm encounters, and broken heads. [The antagonists] were the friends of the person executed ... and persons sent by private surgeons to obtain bodies for dissection. The contests between these were fierce and bloody, and frightful to look at.

Scenes such as these, and the growing opprobrium surrounding anatomization, damaged the image of the establishment in general and the medical profession in particular. In John Gay's 1728 *Beggar's Opera*, the character Matt of the Mint anathematizes the physicians:

> Poor brother Tom had an accident this time twelve-month, and so clever a made fellow he was, that I could not save him from those flaying rascals the surgeons; and now, poor man, he is among the otamies [anatomists] at Surgeon's Hall.

Anatomists had become associated with criminals and worse. Even legal dissections seemed to transgress against religion and humanity, but the constant shortage of cadavers meant that medical students, anatomists and their agents were forced to become "resurrection men": grave robbers who exhumed bodies or raided mortuaries. Public revulsion could spill

EACH INDIVIDUAL'S FLESH

In the Christian West, with its doctrine of resurrection at the time of the Last Judgement, disinterment and anatomization had profound religious implications. The church fathers had insisted the unity of soul and body to be essential to the survival of an individual except as a sort of ghost, and this was amplified by church doctrine on the technicalities of the resurrection. Catholic and many Protestant catechisms specified that it would be the "particles composing each individual's flesh" that would be used to reconstitute "the identical structure, which death had previously destroyed". In this reading, the corpse of the deceased must be kept together; even if it decomposed entirely, the "particles composing each individual's flesh" would be on hand to be raised from the dead. Accordingly, the dissection and dismemberment of an individual's flesh occasioned by anatomization, which might well include specimens separated and bottled, not only outraged human dignity but threatened the victim's prospect of eternal life.

over into rioting (*see* box on page 128), while real and imagined fear of the resurrection men prompted the marketing of "anti-snatching" technology such as fortified crypts and mausoleums, or the leaden patent coffin. This device, registered in 1817, the year of *Frankenstein*'s composition, was advertised as "the only remedy" against the "depredations" of the body-snatchers, with alarming rhetoric about "the well-known fact that many hundreds of Graves and Vaults are constantly disturbed".

The logical solution to the unmet demand for cadavers was to increase production by fair means or foul, and the Burke and Hare affair of 1828 (*see* box on page 130) would have been fresh in the minds of the audience for the 1831 third edition of Mary Shelley's novel. The horrors of "Burkeism", especially the generally assumed complicity of Dr Robert Knox, dramatically changed how the public would have read the text, and even given a grim resonance to the cameo made in the novel by the city of Edinburgh, which Victor and Clerval visit en route to the Orkneys in Chapter 19.

THE BONE BILL

Since ancient Egyptian times the sanctity of the interred body was stoutly defended, and the activities of grave robbers or others who would disturb the dead were regarded with hatred and revulsion. William Beckford's 1786 Gothic novel *Vathek* helped introduce into English the term "ghoul", derived from the Arabic *ghul*, a graveyard-haunting, corpse-eating fiend. Similar loathing applied to those who procured and sometimes purloined corpses for the anatomist's table, and public antipathy could spill over into violence. One of the most extreme examples was the New York "Doctors' Riot" of April 1788, when a man discovered that one of the corpses being dissected by medical students at the Society of the Hospital of the City of New York was that of his wife, robbed from her freshly dug grave. In a letter to Edmund Randolph, the Governor of Virginia, Colonel William Heth described what happened next:

> The cry of barbarity and etc. was soon spread—the young sons of Galen [i.e. the medical students] fled in every direction— one took refuge in a chimney—the mob raised—and the Hospital apartments were ransacked. In the Anatomy room, were found three fresh bodies—one, boiling in a kettle, and two others cutting up—with certain parts of the two sex's hanging up in a most brutal position. The circumstances, together with the wanton and apparent inhuman complexion of the room, exasperated the Mob beyond all bounds, to the total destruction of every anatomy in the hospital.

The 5,000-strong mob rioted for three days at the cost of twenty lives, being dispersed only when the militia opened fire. A year later the New York legislature passed "An Act to Prevent the Odious Practice of Digging Up and Removing for the Purpose of Dissection, Dead Bodies Interred in Cemeteries or Burial Places", popularly known as the "Bone Bill".

*Physicians and students engaged in a dissection
are visited by the mob; public revulsion at the
anatomist's trade could spill over into riots.*

BURKEING GHOUL

In 1827 William Burke and William Hare, two Irish labourers living in Edinburgh, Scotland, decided to sell the body of a man who had died in their lodging house, and who owed them money. Receiving the significant sum of £7 from a Dr Robert Knox, who ran an anatomy school, they realized they had stumbled onto a lucrative profession and subsequently murdered sixteen people to supply as cadavers. Burke, at least according to testimony provided by Hare, who in turn avoided the hangman's noose, perfected a method of murder by asphyxiation that left few trauma marks, which came to be known as "burkeing" (recall that Frankenstein's monster throttles his victims). When the murders were discovered in 1828, it was widely assumed that Dr Knox was complicit, but he refused to comment and opinions differ as to whether he was aware of the provenance of the corpses. Burke was hanged in 1829 and dissected by a rival of Dr Knox. Similar cases were discovered in London. The London burkers, John Head and John Bishop, were dissected at the Royal College of Surgeons in 1831, as was Elizabeth Ross in 1832, the only convicted female "Burkeite".

RESTORED TO LIFE: RESUSCITATION AND THE TENUOUS BOUNDARY BETWEEN LIFE AND DEATH

When Victor Frankenstein discovers the secret of life and the key to reanimating the dead, the loftiest dreams are vouchsafed him. He would become the first to break the bounds of life and death, he boasts, "and pour a torrent of light into our dark world. A new species would bless me as its creator and source ... No father could claim the gratitude of his child so completely as I should deserve theirs." It does not occur to him that the object of his exultations might neither bless him nor be grateful. Yet his author, Mary Shelley, knew all too well that not every body brought back from the dead would bless their reviver or be grateful to be revived; her own mother, Mary Wollstonecraft, had been the unhappy recipient of resuscitation.

Resuscitation was a matter of the first moment in the era of Frankenstein. It was, says Carolyn Williams, a professor of English at Rutgers University, "a richly documented, highly conspicuous area of scientific endeavour, which generated much excitement in life and literature from the last quarter of the eighteenth century onwards". Mary Shelley lived at a time of remarkable advances that both promised to permeate the boundary between life and death and provoked anxieties over its geography. How could doctors tell if someone were truly dead, when medical science continuously improved? Could advances in scientific and medical knowledge be recruited in the war against death? Fears over premature burial mingled with the keenest hopes for medical breakthroughs and wonder at exciting new technologies. All these would influence the young writer's fancy, filtered through the lens of painful and tragic personal history.

THE CALMEST ACT OF REASON

Though driven to murderous rage by abuse and cruel rejection, the monster nonetheless has heroic impulses. In a notable scene in Chapter 16 he spies a girl at the river's edge:

> *when suddenly her foot slipt, and she fell into the rapid stream. I rushed from my hiding place, and, with extreme labour from the force of the current, saved her, and dragged her to shore. She was senseless; and I endeavoured, by every means in my power, to restore animation ...*

This scene bears remarkable similarity to one from Mary Shelley's family history. In October 1796, her mother, Mary Wollstonecraft, betrayed and abandoned by her lover, conceived "a fixed determination to die". In what she insisted was "one of the calmest acts of reason", she threw herself into the Thames from Putney Bridge, but was spotted by some boatmen, who fished her out and managed to resuscitate her. Mary was distraught: she had, she complained, using language that almost seems to allude to grave robbers, been "snatched from death" and "inhumanly brought back to life and misery". She felt sentenced to "a living death". This was not the only suicide attempt in Mary Shelley's life: Percy's first wife, Harriet, had drowned herself in the Serpentine, a small river in London's Hyde Park, in November 1816. She was found to have been pregnant at the time. A month before this her step-sister, Fanny Imlay, had taken her own life with laudanum (a morphine extract).

Death attended Mary at closer quarters still; her mother had died from complications in her own birth. Thoughts of this must have been prompted by her own experiences in childbirth (her and Percy's third child, Clara, was born in September 1817, meaning that she was pregnant for most of the period she was writing Frankenstein), and by the national tragedy that occurred while Mary was completing her novel. In November 1817, Princess Charlotte, the only child of George IV, died in childbirth along with her baby. The British people had been excitedly expecting a long-awaited heir to the throne; now the succession was thrown into doubt and the nation grieved for a popular young royal, just twenty-one years old. Charlotte's labour had dragged on for fifty hours and it is possible that attempts to hasten the delivery contributed to her death. The royal obstetrician, Sir Richard Croft, committed suicide the following year.

Other deaths, yet more painful, haunted Mary. She had given birth to her first child with Percy in February 1815, but the infant girl, born two months premature, did not live long. "Tis hard indeed for a mother to loose [sic] a child," she wrote in her journal, admitting that she could not shake thoughts of "my little dead baby" and that "whenever I am left alone to my own thoughts & do not read to divert them they always come back to the same point – that I was a mother & am so no longer". On 19 March 1815 she recorded:

> *Dream that my little baby came to life again – that it had only been cold & that we rubbed it by the fire & it lived – I awake & find no baby – I think about the little thing all day.*

A portrait of Mary Wollstonecraft, from the year of her death, shortly after giving birth to her daughter Mary Godwin.

What if it were possible to bring someone back from the brink of death? In 1814 Mary is believed to have read accounts of a celebrated case in which a sailor was revived after lying in a coma for many months. The physician involved in the case was Henry Cline – she had once been his patient. What if it were even possible to bring someone back from beyond the threshold?

REKINDLING THE EXPIRING SPARK

Mary Shelley was not the only person to have considered this question. In London in 1774, the "Society for the Recovery of Persons Apparently Drowned" was founded (it would change its name to the Royal Humane Society (RHS)). It was established in response to the escalating incidence of accidental and deliberate drownings in London's many rivers and canals. Increasing numbers of people were living and working by the water, but few knew how to swim, while the unrelenting pressures of poverty and overpopulation took a grim toll on psychological health and led to an epidemic of suicides. The RHS sought to research and disseminate information on resuscitation, and reward rescue attempts (successful rescues attracted greater rewards). People restored to life, including thwarted suicides, would march in an annual procession to celebrate the RHS and its methods. Some of these methods could be alarming. One recommendation was to blow smoke into the anus of the insensible person; another was to stuff electrodes into the subject's orifices and apply an electrical shock (*see* page 52 for more on early claims of defibrillation).

A more refined version of this electroshock therapy was promised by Giovanni Aldini, nephew of Galvani, as the true motivation behind his gruesome spectacle of January 1803, in which the corpse of the executed murderer George Forster was subjected to galvanic reanimation (*see* page 54). According to the account in the *Times* of London, the object of the demonstration "was to shew the excitability of the human frame, when this animal electricity is duly applied. In cases of drowning or suffocation, it promises to be of the utmost use, by reviving the action of the lungs, and thereby rekindling the expiring spark of vitality."

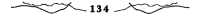

In his *General Views on the Application of Galvanism*, Aldini himself inveighed against the defeatism of contemporary medicos:

> *Numerous instances could be produced in which persons have been hurried to the grave before life was entirely extinct. I view with horror and indignation the haste with which a man, who appears to have drawn his last breath, is thus banished from society, and deprived of a chance of recovery.*

One remedy, Aldini proposed, might be to equip armies of children with portable galvanic machines "weighing from 24 to 38 ounces", so that they could "be taught its value from their tenderest years and afterwards learn to apply it in cases of suspended life". In some senses far-sighted – given modern programmes to equip all schools and similar institutions with easy-to-operate defibrillators that anyone can use – Aldini must nonetheless be viewed with suspicion here, given that he presumably imagined himself supplying the devices at considerable cost.

A LOATHSOME OCCUPATION

New technologies such as electroshock devices or blood transfusion (*see* box on page 137), and astounding demonstrations such as Aldini's galvanic reanimation and Waterton's ass (*see* box on page 138), meant that the boundary between life and death was increasingly contested and blurred. In this strange new world, how were physicians to know that life was "entirely extinct"? Inaccurate determination of death raised hideous spectres: premature interment, corpses coming back to life, people buried alive.

The late eighteenth century saw a spike in reports of premature burials, and while it is not clear how many were genuine, the fear they engendered was very real. George Washington, for instance, with his last words, instructed his doctor: "Have me decently buried, but do not let my body be put into a vault in less than three days after I am dead." Such anxieties fed demand for technologies that could assuage them. A number of "safety coffins" were designed, typically featuring connections to bells that could be sounded above ground, so that the person in the coffin, on regaining consciousness, could pull a chord and alert the outside world to their predicament. Whether or not anyone was ever buried, let alone saved, in such a device, post-burial interventions were of less interest to physicians than actually being able to declare death with certainty. The composer Frédéric Chopin's last words alluded to these: "The earth is suffocating ... Swear to make them cut me open, so that I won't be buried alive."

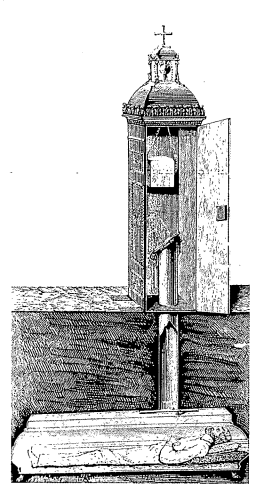

A late Victorian patent safety coffin, descendant of
related devices that began to appear in the late eighteenth
century to exploit anxieties about grave robbing.

BLOOD TRANSFUSION

In the same year that *Frankenstein* was published, there was a major advance in emergency medical care for women haemorrhaging after childbirth (a condition that would nearly kill Mary Shelley after she miscarried in June 1822). Blood transfusion had been attempted ever since William Hervey had demonstrated circulation of the blood, but most early attempts had involved animal-human transfusions, which were so dangerous that they were banned. The first successful human-human transfusion probably dates to 1795, when the aptly named American physician Philip Syng Physick performed one in Philadelphia, but this was not reported. In 1818, British obstetrician James Blundell became the first to perform a successful transfusion to treat post-partum haemorrhage, using the patient's husband as a donor. Blundell extracted four ounces (~110 ml) of blood from the husband's arm and used a syringe to inject it into his wife. He would go on to perform the technique several times, with success in about half of cases (it is now known that blood types must be matched or transfused blood will be rejected by the recipient's immune system, with potentially dangerous results).

Tests to ascertain clinical death were a concern of the anatomists at the Royal College of Surgeons, where it was claimed that some of the "corpses" of hanged people had actually revived on the College's own dissecting table. Accordingly, murderers' bodies were used for experiments on how to prove death, and these veered into territory so macabre and gruesome that even the College's own professional anatomist was distressed. William Clift, Conservator of the College's Hunterian Museum, whose job it was to perform the dissections, kept a note book detailing his work. His handwritten record of trials performed on the body of Martin Hogan in 1814 suggest extreme stress and discomfort, with crossed out words, redrafting and shaky penmanship: "[a] needle was immediately introduced ~~immediately~~ through the coats of ~~the~~ each eye ~~& each? With the view of stimulating~~ to the Iris – but no visible effect was produced." One is reminded of Victor Frankenstein's account of the stress of his own "secret toil": "often did my human nature turn with loathing from my occupation".

WATERTON'S ASS

The Georgian public was treated to some extraordinary demonstrations of the potential power of science to control or conquer death. The grisly spectacles of Aldini and Ure (*see* pages 54 to 59) employed electricity, while the English naturalist and explorer Charles Waterton used a paralysing drug he had brought back from the Amazon. In 1814 Waterton returned to England with samples of curare, a poison made from the bark of a tree, which South American indigenous peoples used as an arrow or dart poison. Curare (from the Amazonian native word *uirary,* meaning "bird killer") is a paralytic agent, which stops animals from breathing so that they suffocate. In 1811, English physician Benjamin Brodie had reported to the Royal Society that artificial inflation of the lungs of rabbits and cats poisoned with curare (which he called woorara) would sustain their heart function and keep them alive until the poison wore off. Waterton had probably learned of Brodie's experiment, and went to great lengths to acquire a quiver of arrows tipped with the finest quality curare (although he transliterated the indigenous term as wouralia) from deep in the Amazon, in the south of Guiana near the border of Brazil.

Back in England Waterton was invited to the Royal Veterinary College, to experiment on asses. One ass was inoculated with the poison and died after twelve minutes. A second was inoculated below a tourniquet and only died when the binding was released, allowing the poison into the main circulation. A third ass was then inoculated in the shoulder. Waterton described what happened next:

A she-ass received the wourali poison in the shoulder and died apparently in ten minutes. An incision was made in its windpipe, and through it the lungs were regularly inflated for two hours with a pair of bellows. Suspended animation returned.

The ass held up her head and looked around; but the inflating being discontinued, she sunk once more in apparent death. The artificial breathing was immediately recommenced and continued without intermission for two hours more. This saved the ass from final dissolution; she rose up, and walked about; she seemed neither in agitation nor in pain.

The ass eventually recovered her full strength, was re-named Wouralia, and sent to live "in the finest pasture" to "end her days in peace".

CHAPTER 6

TO THE ENDS OF THE EARTH

POLAR EXPLORATION AND THE FROZEN WASTES

THE ROMANTIC ERA OF SCIENCE WAS MARKED BY AN ASSAULT ON THE LIMITS OF GEOGRAPHICAL KNOWLEDGE, AS BOLD EXPLORERS EXPANDED THE HORIZONS OF IMAGINATION AND LITERATURE. WHAT DEBT IS OWED TO THEIR GRAND AMBITIONS AND DELUDED SCHEMES, AND TO THE REAL SCIENCE OF POLAR DISCOVERY, BY THE NOVEL'S BLEAK VISION OF THE POLAR WASTES?

QUEST FOR THE NORTHWEST PASSAGE

Those with a casual acquaintance with the tale of *Frankenstein* are often surprised to learn that the book begins and ends in the frozen wastes of the Arctic. In a framing device not unusual in fiction of the period, an Arctic explorer named Robert Walton writes a series of letters to his sister. The first paragraphs of the novel recount his ambitions and mounting excitement as he prepares in St Petersburg for an exploration to the uttermost north, followed in short order by his arrival in the frozen seas. Once there, Walton and his crew are first startled to see a monstrous form driving a dog sled past them, and then amazed to pull alongside an enfeebled Victor Frankenstein, who is pursuing his creation. It is Walton to whom the tragic tale of Frankenstein is related, and it is he who delivers the coda to the story, detailing Victor's death and the monster's final visit to his creator.

The bleak and terrible landscape of the frozen north mirrors the psychological and spiritual torment of the protagonists; a classic device of Romantic narrative. But it also reflected with extreme immediacy and extraordinary prescience the very latest in the science and politics of polar expedition. What was the background to this polar narrative, what were the influences on Mary Shelley's invention and how did the fictional expedition relate to reality?

THE ICE WAS ALL AROUND

The icy regions of the world held a special fascination for the Romantic imagination. The Shelleys and Byron – like Coleridge and Wordsworth before them – found themselves awestruck by the sublimity of Alpine landscapes and glaciers. But they were also appalled by the bleak desolation that icy vistas could present – at least, certainly in the case of Mary Shelley. Such fascinations extended, at greater remove, to the poles, which were among the last undiscovered frontiers of the world in the early nineteenth century.

Since her teenage years, Mary had been enthralled by accounts of early Arctic voyages, which she records that she "read with ardour".

It is possible that her interest was inspired by Coleridge himself; as a child she had sneaked downstairs in the evening to listen to the great poet recite his *Rime of the Ancient Mariner*, in which the eponymous seaman paints an evocative picture of the polar seas:

> *The ice was here, the ice was there,*
> *The ice was all around:*
> *It cracked and growled, and roared and howled,*
> *Like noises in a swound!*

However, the polar section of *Frankenstein*'s narrative was added to the novel relatively late in the day; its inclusion reflected current events with which Mary was almost certainly familiar.

Early nineteenth-century depiction of sailing in the polar waters; the Gothic contortions of icebergs contributed to the Romantic appeal of the northern realms.

The press was full of talk of polar expeditions – particularly the efforts of John Barrow, Second Secretary of the Admiralty, to push for an Admiralty-sponsored search for the Northwest Passage, the fabled sea route between the Atlantic and Pacific Oceans around the top of North America. It is this quest to which Walton refers when he dreams of "the inestimable benefit which I shall confer on all mankind, to the last generation, by discovering a passage near the pole."

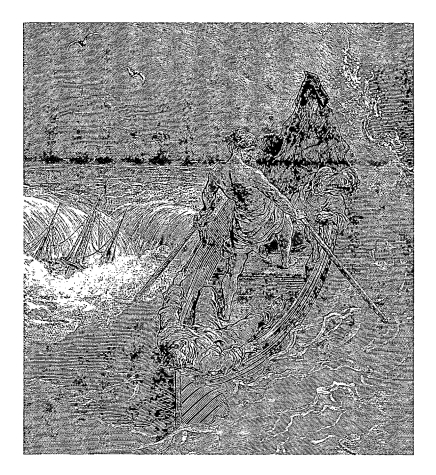

Engraving in the style of Gustave Doré, from an illustrated edition of Coleridge's Rime of the Ancient Mariner, *whose eponymous protagonist encounters polar wonders.*

PASSAGE TO NOWHERE

The Northwest Passage was the polar El Dorado – a mirage that, for centuries, drove explorers to the edge of reason and beyond the bounds of life. While geographers had known since the time of the ancient Greek explorer Pytheas (c.330 BCE) that the far north was a place of frozen seas, where man could "neither sail nor walk", it was assumed by early modern mariners that the sea route from Europe west to the Indies lay open, and this was precisely the motivation of Columbus when he set out to sail the ocean blue in 1492. Transatlantic explorers discovered that the Americas lay in the way, and although Magellan and others successfully sailed around the bottom of South America, this Atlantic–Pacific route was distant, dangerous and, for northern European nations, inhospitable, due to Iberian control of the south of the New World. Similar objections applied to the circum-African route to the Indies, so that northern mariners were motivated to find a passage from the northwest Atlantic into the Pacific. John and Sebastian Cabot, sailing out of England, became the first Europeans since the Norsemen to set foot on the North American continent when they visited in the late fifteenth century. Their reports made it clear that any putative Northwest Passage would involve travelling far to the north around the American landmass. It was believed that a strait known as Anian existed, and late sixteenth- and early seventeenth-century attempts to find it led to the discovery of many features of the North American Arctic, named after the explorers who sought the Northwest Passage: Frobisher, Hudson, Davis and Baffin.

These efforts, however, often came with a terrible price. Henry Hudson, for instance, suffered a mutiny in 1611 when provisions ran out as he attempted to find the Passage; he was set adrift in a boat with eight companions and never seen again. Two expeditions in 1631 that suffered scurvy and horrible winter conditions were so traumatic that it would be nearly a century before further voyages were launched. In 1719, the Hudson Bay Company veteran James Knight launched an assault on the Northwest Passage, but his two ships vanished without trace. Further expeditions by sea and by land could find no way through the maze of islands and ice-choked waterways, and even Captain James Cook was turned back by pack ice when he attempted to chart a passage from the Pacific to the Atlantic.

By the late eighteenth century some experts had come to believe the Northwest Passage to be a delusion. Alexander Mackenzie, the first white man to cross North America, twelve years before Lewis and Clark, concluded that the Passage did not exist, and suggested that resources might be better directed to developing a land route:

> The discovery of a passage by sea, North-East or North-West from the Atlantic to the Pacific Ocean, has for many years excited the attention of governments, and encouraged the enterprising spirit of individuals. The non-existence, however, of any such practical passage being at length determined, the practicability of a passage through the continents of Asia and America becomes an object of consideration ...

Similarly, George Vancouver, a veteran of Cook's second and third expeditions, explored the northwest coast of North America over three years of painstaking charting. In 1798 he declared that he had dispelled "every doubt" about the existence of a passage "between the North Pacific, and the interior of the American continent, within the limit of our researches". It did not exist.

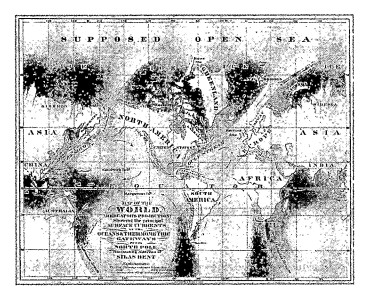

Silas Bent's 1872 map showing the Open Polar Sea and the "thermometric gateways" that led to it.

BARROW BOYS

Nevertheless the dream persisted, fired by the resurgent theory of the Open Polar Ocean (*see* page 150), and by 1817 a new quest for the Northwest Passage was the talk of the town. John Barrow, whose incessant boosting tapped into the spirit of the time, relentlessly promoted such an expedition. Britain's large and restless navy – now conspicuously underemployed in the aftermath of the Napoleonic Wars – sought new goals and fresh glory, while Barrow stoked imperial pretensions with a series of stirring articles in the *Quarterly Review*. In these he warned of the perils of Russian ambition and their imminent conquest of the Arctic, and even implicitly challenged the pride and virility of the nation. Should Britain fail to find the Northwest Passage, he said, the country "would be laughed at by all the world."

It was against this macho, imperialistic backdrop that Mary Shelley very deliberately crafted her fictional version of a "Barrovian" explorer. She was almost certainly aware of John Barrow and the wider background of polar exploration. Critic Jessica Richard has established that Mary probably read Barrow's *Quarterly Review* articles, while Percy and Mary Shelley's own record of their reading list for November 1816 states that they "read old voyages". It would have been clear enough from press debates on the topic that arctic exploration was not nearly as straightforward as Barrow was claiming; he played up the "scientific" evidence in favour of his Open Polar Ocean hypothesis and downplayed the dangers and costs of such expeditions. Mary perhaps saw parallels between the hubristic ambition of her protagonist and the real-life Dr Frankensteins willing to risk men's lives in vainglorious pursuits of dubious scientific merit; her choice of framing device was no accident.

The character and conduct of Walton both mirrors that of Victor Frankenstein and serves as a veiled commentary on the real adventurers. Like Victor, Walton is motivated by grandiose ambitions and scientific arrogance, boasting of "the inestimable benefit that I shall confer on all mankind, to the last generation", and like Victor his hubris leads to nemesis. By the end of the novel he writes despairingly: "We are still surrounded by mountains of ice, still in imminent danger of being crushed in their conflict.

The cold is excessive, and many of my unfortunate comrades have already found a grave amidst this scene of desolation."

Indeed, Shelley's construction of Walton can be read as depicting an incompetent and intellectually dishonest failure of an explorer. He sets off at the wrong time of year, from the wrong place for one wishing to make fresh discoveries. His scientific goals (*see* page 153) seem sketchy and his commentary on events and the environment does not project competence. The polar narrative section of the novel can thus be read as a deliberate critique of imperial pretensions and the purpose of Arctic exploration. According to Jessica Richard, "far from simply appropriating a contemporary discussion uncritically, [Shelley] must be counted among [the] voices that censured the revival of British polar exploration".

Cruikshank's 1819 cartoon "Landing the Treasures, or Results of the Polar Expedition!!!" satirized the utility and success of the 1818 expeditions.

BLASTED HOPES

In the final chapter of *Frankenstein*, Walton's vessel lies "immured in ice" and "in imminent danger of being crushed". His crew entreats him not to risk their lives by seeking to go on, should the ship be freed and a passage home open up. Walton relents: "I have consented to return if we are not destroyed. Thus are my hopes blasted by cowardice and indecision; I come back ignorant and disappointed." Sure enough, the ship is released from the grip of the ice and a despondent Walton retreats from the Arctic, even as Victor Frankenstein draws his last breath.

Walton's narrowly avoided catastrophe grimly prefigures the disaster that would befall Sir John Franklin in his ill-fated 1845 expedition, the "Moon shot" of its day. Franklin set off equipped with the height of mid-nineteenth-century technology, but was lost after his ships were trapped in pack ice. His fate remained a mystery, prompting multiple, often fatal search expeditions in a saga that gripped the British public for decades. More immediately, Walton's ignominious retreat uncannily foretold the progress of an expedition launched just months after its publication. John Ross's Expedition of 1818 penetrated the Canadian Arctic as far as Lancaster Sound, but Ross turned back after claiming he could see "a continuity of ice" and mountains blocking the way ahead. His reputation was destroyed as a result; it seems the unforgiving British public did not share Mary's apparent sympathies for those who considered discretion to be the better part of valour, and would rather live than achieve a "glorious death" in the frozen north.

THE OPEN POLAR SEA

Mary Shelley's hapless polar explorer Robert Walton espouses a geographical doctrine that may seem strange to modern conceptions of the Arctic, although ironically it comes nearer to reality every summer thanks to climate change. His expedition is founded on the belief that beyond the belt of pack ice, the frozen sea where man can "neither sail nor walk", lies an ice-free realm of open seas and balmy climes: the open polar sea. The legend of the open pole had deep historical antecedents, but in Shelley's day it had been given a new, "scientific" lease of life.

A LAND SURPASSING IN WONDERS
Writing to his sister in the first chapter of the novel, Walton lays out the rationale for his expedition to the furthest north:

> *I will put some trust in preceding navigators – there snow and frost are banished; and, sailing over a calm sea, we may be wafted to a land surpassing in wonders and in beauty every region hitherto discovered on the habitable globe.*

Walton's hyperbolic language is fitting, because the land to which he refers sounds very much like Hyperborea, the country "beyond the North Wind", which the ancient Greeks believed lay in the uttermost north of the world. It was a fantastic realm of warmth and plenty in which dwelt a superior race. According to the fifth century BCE poet Pindar: "neither ship nor marching feet may find the wondrous way to the gatherings of the Hyperborean people ... Illness and wasting old age visit not this hallowed race, but far from toil and battle they dwell secure from fate's remorseless vengeance."

Diodorus Siculus, Greek historian of the first-century BCE, wrote that "the island [of Hyperborea] is both fertile and productive of every crop, and since it has an unusually temperate climate it produces two harvests each year", while Pliny the Elder, the first century CE Roman encyclopaedist, described how:

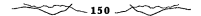

[beyond a] part of the world that lies under the condemnation of nature and is plunged in dense darkness, and occupied only by the work of frost ... there dwells – if we can believe it – a happy race of people called the Hyperboreans, who live to extreme old age and are famous for legendary marvels. Here are believed to be the hinges on which the firmament turns and the extreme revolutions of the star ... for these people the sun rises once in the year, at midsummer, and sets once, at midwinter. It is a genial region, with a delightful climate and exempt from every harmful blast.

Mary, probably well versed in the classics, would surely have heard of the legend, but she would also have been aware of its modern

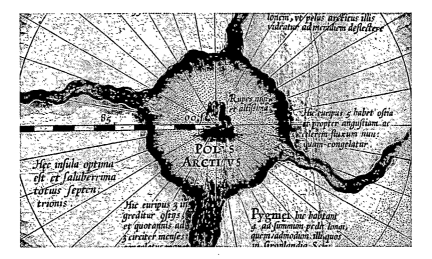

Detail of the North Pole from Mercator's Atlas, showing the mountain of lodestone in the midst of the open polar sea, surrounded by legendary lands such as Hyperborea.

FROM HYPERBOREA TO CROCKER LAND

Hyperborea and the related land of Thule – another island in the far north mentioned by ancient geographers – were the two Atlantises of the Far North. These legendary places provided a template on to which could be projected all manner of fantasies – from gateways to the Hollow Earth to Nazi flying-saucer bases. New additions to the canon of legendary northern lands appeared as late as 1906, when American polar explorer Robert Peary claimed to have sighted a new continent near the pole, which he named Crocker Land after George Crocker, a banker who had backed his expedition. In fact Peary, who was later shown to be a serial fraud, invented this mysterious country out of thin air.

incarnation. From the ancients the legend of a temperate realm beyond the ice had come down to the early modern era. Geraldus Mercator, the influential sixteenth-century German-Flemish cartographer, depicted Hyperborea near the top of the world in his map of the Arctic. He also, perhaps logically, concluded that a balmy clime would mean an ice-free sea, showing an open ocean at the pole itself, and this belief persisted into the mid-nineteenth century and beyond.

THE CASE FOR THE OPEN POLE

By the time Mary came to write *Frankenstein*, the "open pole" hypothesis had accrued various scientific buttresses. Extrapolations of the correlation between temperature and latitude led some to the conclusion that the coldest temperatures occurred at around 80° north, while it was assumed that the continual summer sun beyond the Arctic Circle would warm the pole. Studies of ocean currents suggested that perhaps the Gulf Stream and other warming ocean currents might surface in the polar sea, and it was also believed that pack ice could only form near coastlines, and thus would not be present at the landless pole. It was even believed that the declivity of the Earth's surface at the poles (the Earth is not a perfect sphere, but is slightly flattened at either end) meant that the pole would be nearer the molten core of the planet and thus warmed by geothermal heat.

English Arctic explorer William Scoresby was a champion of the open pole theory and, despite widespread scepticism, some of his writings were recycled by John Barrow at the Admiralty. Barrow ignored any evidence to the contrary, and even distorted Scoresby's analysis, omitting his suggestion that a passage through the encircling ice might only open once every few years. Shelley may well have been aware of Barrow's unreliable presentation of the theory – Walton's adherence to the open pole hypothesis can thus be read as a satirical commentary on this popular delusion.

The constant failure of expeditions to penetrate beyond the pack ice failed to kill off the open pole theory. As late as 1872, maps showing the open pole were still being produced; since no one had yet reached the pole, who was to say what lay there? Reaching the North Pole turned out to be one of the hardest achievements in terrestrial exploration; it was only definitively achieved in 1968!

THE SECRET OF THE MAGNET: TERRESTRIAL MAGNETISM AND THE POLES

Mary Shelley's polar explorer Robert Walton mirrors Victor Frankenstein both in his reckless and overweening ambition to cross the boundaries of knowledge, and in his specific interest in the force we know today as electromagnetism. Today electricity and magnetism are understood to be a single force – electromagnetism – but in Mary Shelley's day this had not been proven. It was, however, strongly suspected, because of the characteristics the two forces shared; not least their mystery. Where Mary gives Victor, through implication at any rate, a fascination

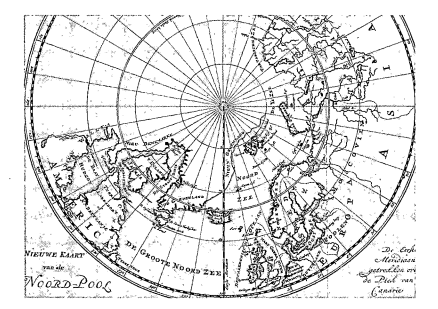

This early eighteenth-century map of the North Pole avoids speculation but reveals just how much of a terra incognita *the pole remained.*

with electricity, she grants Walton a parallel interest in magnetism. Writing to his sister of his scientific aspirations for the expedition to the pole, Walton breathlessly avers, "I may there discover the wondrous power which attracts the needle ... or ascertain the secret of the magnet". Why is Walton seeking the secrets of the magnet at the North Pole, and what was the real science that inspired Mary Shelley to include this in her novel?

THE LODESTONE MOUNTAIN

The compass was invented in China as early as the third century BCE, and it was in widespread use for navigation by Europeans by the twelfth century. A compass is a piece of magnetized metal that is able to rotate freely on an axis or float on a liquid medium, so that it can re-orient itself as it is moved about. What makes it useful is that it will, generally, maintain its orientation along a north-south axis. Medieval Europeans seeking to explain the action of the compass related it to the source of the needles themselves: lodestone, a naturally magnetized iron mineral. Lodestone magnets can attract one another, and so it was surmised that compass needles must have been responding to the influence of a great mass of lodestone, presumably situated in the north: a mountain of lodestone at the North Pole.

Lodestone mountains had been discussed by classical authors such as Pliny the Elder, albeit the ancients located them in India, so the idea of a lodestone mountain was well accepted. The advent of the compass shifted its location to the North Pole, and belief in this Hyperborean magnetic mountain was well enough known for it to feature as a simile for the power of love in Guido Guinizelli's 1276 poem "Madonna, il fino amor ched eo vo porto,":

> In that land beneath the North Wind
> Are the magnetic mountains,
> Which transmit to the air their power
> To attract the iron, but because it is far away,
> It needs help from a similar stone
> To make the compass needle
> Turn towards the pole star ...

Not all authorities accepted the northern lodestone mountain theory, but influential sixteenth-century believers included Girolamo Fracastorio and Olaus Magnus. In his 1555 *Description of the Northern Peoples*, Olaus wrote that northern shipwrights had to use wooden pegs in place of iron nails, which would be wrenched loose by the powerful attraction of the northern lodestone. The most famous cartographer of all, Geraldus Mercator, included in his 1569 map of the Arctic a striking black lodestone mountain at the very apex of the globe, in the centre of the open polar sea (*see* page 151). It is labelled *Rupus Nigra et Altissima*, or "Black, Very High Cliff". Mercator's Arctic map was enlarged to its most familiar form by his successor Jodocus Hondius in 1606 and, perhaps showing that mariners had already realized that the magnetic north to which compass needles pointed did not align precisely with the North Pole, this map depicts another mountain nearer the top of the map, labelled "Magnetic Pole".

GILBERT'S LITTLE EARTH

By the time of the publication of the revised Mercator Arctic map, the lodestone mountain theory had already been fatally wounded. Sir William Gilbert (1544–1603), physician to Queen Elizabeth I, was one of a new breed of natural philosophers; he was sceptical of the ancient authorities that had been enshrined as dogma by the Catholic Church, and open to the practice of experiments and the insights of artificers. Gilbert was particularly intrigued by magnetism – a curious force that could not easily be accommodated in the prevailing Aristotelian physics, but one that was amenable to experimental investigation.

Gilbert probably knew the *Epistola* of the thirteenth-century French scholar Peter Peregrinus de Maricourt, who in 1269 had written one of the earliest western works on the compass and on magnetism. It was from him that Gilbert borrowed the idea of *terrellae* – "little earths". These were lumps of magnetite lodestone, chiselled into spheres, which were intended to represent the Earth in microcosm. Gilbert's genius was to encompass the analogous nature of the *terrellae* as models for the real Earth. Gilbert knew from his contemporary, the compass-maker Robert Norman, that the magnetic force to which compass needles responded acts not only in the horizontal plane but also in the vertical. A compass needle does not simply rotate to point north – it also declines more in some places than others, a phenomenon known as "magnetic dip".

Using a compass-like instrument named a Versorium (from the Latin for "turning around"), Gilbert was able to show that he could trace the contours of a force field around a *terrella*. Near the equator of the *terrella*, the Versorium needle lay flat, parallel to the surface of the sphere and aligned north–south. As the Versorium was brought closer to the pole of the *terrella*, its needle dipped until, at the pole, it pointed straight down. Evidently the lines of force in the magnetic field of the *terrella* described arches, emanating from the poles, and this accounted for magnetic dip. Gilbert believed that this curved field of magnetic force derived from the "soul" of the Earth itself and, he concluded in his great work *De Magnete* (often considered to be the first systematic scientific study): *magnus magnes ipse est globus terrestris,* "the earth's globe itself is a great magnet."

In his detailed study, Gilbert also showed that irregularities in the surface and composition of the *terrella* affected the movement of the Versorium needle, corresponding to similar irregularities that were displayed by real-world compass needles, which he concluded must be due to local variations in the Earth's magnetic field.

As Mercator or his school apparently already knew, navigators in the northern seas noted discrepancies between the target of their compass needles, magnetic north, and the True North, as determined by the intersection of the lines of longitude. This had important consequences for navigational calculations, and refining precise measurements of the discrepancy may have been among the "magnetic secrets" that Walton hoped to find.

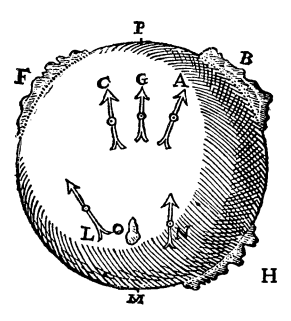

One of Gilbert's terrellae, *complete with miniature mountains made from lumps of iron, which he used to explore the phenomena of magnetism and the compass.*

THE AURORA BOREALIS

Walton also speaks of his ambitions to "regulate ... celestial observations", which could in part be a reference to the mysterious and beguiling celestial phenomenon known as the northern lights, or Aurora Borealis (literally, "northern dawn", so named because the lights can resemble the sun coming up over the horizon). The lights had obvious appeal to the Romantic sensibility; they are caused by the shape of the magnetic field as elucidated by Gilbert with his *terrella*. When charged particles from the Sun approach the Earth, they are captured by and channelled along its magnetic field, following the lines of force until they are led to crash into the atmosphere at the poles. As they interact with the tenuous upper atmosphere, the energetic particles create the strange lights of the Aurora (and the associated noises sometimes heard).

In an intriguing coda to the story of Gilbert and his *terrella*, the Norwegian physicist Kristian Birkeland used a similar *terrella* to prove the nature and origin of the Aurora. Birkeland suspended a magnetic *terrella* in a vacuum chamber and fired beams of electrons at it, demonstrating that the electrons are indeed steered to the magnetic poles by the field.

Physicist Kristian Birkeland (left), with his assistant Olaf Devik, stands next to the vacuum chamber conducting the terrella experiment.

CHAPTER 7

BEFORE AND AFTER

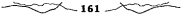

THE PAINFUL BIRTH AND DREADFUL AFTERLIFE OF FRANKENSTEIN AND HIS MONSTER

FRANKENSTEIN BEGAN AS A SHORT STORY, GREW INTO A NOVEL AND EVOLVED INTO A MYTH. THE MAN AND HIS CREATURE WERE CONCEIVED IN DARKNESS (OR GLOOM AT ANY RATE), AND THEIR BIRTH IS SHROUDED IN LEGEND, AS MARY RETROSPECTIVELY EMBELLISHED IT WITH SUITABLY GOTHIC DETAIL, RELATING A WAKING DREAM IN WHICH SHE SAW "WITH SHUT EYES, BUT ACUTE MENTAL VISION ... THE PALE STUDENT OF UNHALLOWED ARTS KNEELING BESIDE THE THING HE HAD PUT TOGETHER ... THE HIDEOUS PHANTASM OF A MAN". THIS PHANTASM HAS GONE ON TO CAST A LONG SHADOW OVER MUCH OF SUBSEQUENT SCIENCE AND TECHNOLOGY, DEVELOPING INTO ONE OF THE MOST POTENT METAPHORS IN MODERN CULTURE.

THE DISMAL SEASON: FRANKENSTEIN AND THE "YEAR WITHOUT A SUMMER"

June, 1816; Mary Wollstonecraft, Percy Shelley, Lord Byron and their friends shiver through an unseasonably damp holiday by Lake Geneva. Mary recalled that it was "cold and rainy", while Byron later captured in verse how the "bright sun was extinguish'd" and the "Morn came and went – and came, and brought no day". This dismal period would become known as the "year without a summer"; for Mary and her friends, forced to stay indoors and entertain themselves with literary parlour games and readings of ghost stories, it would inspire a burst of Gothic creativity unparalleled in literary history, with *Frankenstein* just one among several unholy offspring. Why was the season so dismal? Why did crops fail in New England, Ireland and many other parts of the world, bringing famine and epidemics?

DARKNESS FALLS

Climatologists believe that the long period of unusually low temperatures, high rainfall and low sunshine throughout 1816 was driven by a cataclysmic event that took place on the far side of the world a year earlier: the eruption of Mount Tambora. This eruption was the most violent volcanic eruption of historical times, and the most explosive in the last 10,000 years. In April 1815, Tambora, a 3,962m (13,000ft) high mountain on the Indonesian island of Sumbawa, exploded, hurling around 100 km³ (24 cubic miles) of rock and ash into the sky, vaporizing the communities on the island and sending tsunamis racing outwards to devastate surrounding areas. Of the 12,000 inhabitants of Sumbawa just 26 survived, while the total death toll in the region has been estimated at around 100,000. Within 320km (200 miles) of the volcano there was darkness for three days, as the vast plumes of ash and dust were blasted into the upper atmosphere. Along with aerosols of sulphur and other gases, the ejected material diffused through the upper layers of the troposphere, spreading out around the planet to mask sunlight, lower temperatures, seed rain clouds and increase precipitation.

CLIMATE TURNED BACKWARD

Over the next year or so the effects of this colossal event worked through the global climate system, particularly in the Northern Hemisphere. New England and Western Europe were badly affected and left the most accessible records of what became known as "eighteen hundred and froze to death". While the global temperature average fell by 0.4–0.7°C (0.72–1.26°F), the temperature in these regions fell by up to 2.5°C (4.5°F) below normal. In the northeastern United States, the locals said that the weather in May had turned "backward", while Virginian Pharaoh Chesney wrote that June

Contemporary illustration of the eruption of Mount Tambora volcano in 1815, which would impact the global climate over the following year.

brought "another snowfall and folk went sleighing". He recalled that "On July 4, water froze in cisterns and snow fell again, with Independence Day celebrants moving inside churches where hearth fires warmed things a mite."

Western Europe, not yet recovered from the depredations of the Napoleonic Wars, was badly hit; in Ireland, for instance, eight weeks of constant rain led to a potato famine and subsequent typhus epidemic. Against such real horrors, the inconvenience suffered by a party of genteel holidaymakers in Switzerland seems insignificant, yet it was this that stimulated the most enduring cultural product of the volcanically engendered gloom.

TRACKING DOWN THE REAL FRANKENSTEIN

The enduring resonance of the novel's themes has ensured that Frankenstein has become an iconic name – shorthand for a range of tropes and anxieties around science and human endeavour, and for the archetype of the mad and/or hubristic scientist. The rich stew of scientific and cultural influences for the novel includes a bewildering range of individual inspirations for the figure of Victor, spanning history, geography and literature. Was there a real-life inspiration for the fictional Frankenstein?

PROMETHEAN FIGURES

The subtitle of the novel – "the modern Prometheus" – explicitly names the literary/mythical progenitor of its protagonist. In Greek myth, Prometheus was a Titan who was intimately involved in the fate of humanity, although there were several different versions of his story. In some, Prometheus creates humankind, moulding the first people from clay; in others he steals the secret of fire from the Gods and gives it to humankind, kick-starting civilization and all its arts and artifice. For his presumption, the Titan is punished by Zeus (in some versions betrayed, despite having aided the Gods in their war with the Titans, in order to protect humanity). Prometheus is chained to a rock or mountaintop and sentenced to have his liver pecked out every day for eternity. As a revolutionary figure who dared to defy unjust authority and sought to improve the lot of humanity, Prometheus became a hero to the Romantic movement; he was, for instance, the star of Percy Shelley's epic drama *Prometheus Unbound*. As one who is punished for "inventing" technology that allows humans to master nature, trespassing on the preserve of the Gods, Prometheus is an obvious point of reference for Victor and all the Frankensteins who have come after him.

The doomed Titan may also have helped inspire John Milton's characterization of Satan in his epic poem *Paradise Lost*, another rebel who defies ultimate authority to assert his independence, and is condemned for it. The monster in Mary Shelley's novel explicitly compares himself to Milton's Satan, but Victor too owes something to this anti-hero, another who inspired Romantic sympathies.

Another literary forebear of Victor is Faust, a legendary magus based on a sixteenth-century alchemist, whose story spawned multiple versions – including celebrated plays by Christopher Marlowe, whose *Doctor Faustus* dates to c.1590, and Goethe, whose *Faust, Part One* was published in 1808. The legendary Faust was a scholar who traded his soul to the devil in return for unlimited knowledge and worldly pleasure, at terrible cost. Like Victor, Faust comes to regret his transgressive desire for knowledge; Marlowe, for instance, has him lamenting that he wishes he "had never read book". In some versions of the tale, Faust creates a homunculus of his own, who turns out to be articulate and sensitive. The real-life figure behind the stories is said to have been an itinerant astrologer and alchemist named Johann Georg Faust, who was born in the late fifteenth century and died c. 1541, although whether or not there ever was such a person is now hard to discern.

THREE WISE MEN

Three alchemists who definitely did exist, who are cited in the novel by Victor as inspirations, and who make obvious candidates for the "real Frankenstein", are Albertus Magnus, Cornelius Agrippa and Paracelsus. Albertus Magnus (c. 1200–80) was a German monk and alchemist who taught at the University of Paris, was known as "Doctor Universalis" and later became the patron saint of natural philosophers. Many legends accrued to him after his death, including that he practised magic and had built a brass housekeeper. Mary's father William Godwin, who wrote a multi-volume account of historical and legendary magicians and wonder-workers called *Lives of the Necromancers*, recorded a version of the story with clear relevance to his daughter's creation:

> *It is related of Albertus, that he made an entire man of brass... This man would answer all sorts of questions, and was even employed by its maker as a domestic... this machine is said to have become at length so garrulous, that Thomas Aquinas ... finding himself perpetually disturbed ... by its uncontrollable loquacity, in a rage caught up a hammer, and beat it to pieces. According to other accounts the man of Albertus Magnus was composed, not of metal, but of flesh and bones like other men ...*

Woodcut of Albertus Magnus, aka Doctor Universalis, a Dominican friar said to have created a brazen automaton that grew vexatious.

Cornelius Agrippa (1486–1535) was a German mystic, physician and alchemist. Alchemy stressed the generative powers of putrefying matter (*see* page 78), and Agrippa was said to have devoted much effort to breeding life from dead flesh, a feat in which Victor of course succeeds. Paracelsus, aka Theophrastus Bombastus von Hohenheim (1493–1541), was another German physician and alchemist much concerned with the creation of artificial beings, and is particularly associated with his recipe for a homunculus, or "little man" (*see* page 15).

CASTLE FRANKENSTEIN

Notably all three of Victor's alchemical idols were German, and it is generally assumed, given his name, that Victor himself is German. In fact he is a Swiss from the French-speaking region near Geneva, although this has not stopped him from becoming, in the words of Grant McAllister, an academic specializing in eighteenth- and nineteenth-century literature, the primary source for "a new theatrical Germanic archetype that tended to identify irresponsible but awe-inspiring science and ungodly knowledge within the Anglo-American cultural consciousness as German in nature".

There was, however, a real Frankenstein family – and they were indeed German. Percy and Mary Shelley themselves had sailed past the real-life Castle Frankenstein, in Hesse near Darmstadt, and one popular candidate for the "real Victor Frankenstein" is a shady figure associated with the castle. Johann Konrad Dippel (1673–1734) was a German theologian and alchemist who was apparently born at Castle Frankenstein and hence sometimes signed his name as "of Frankenstein". He is said to have created an *elixir vitae* known as Dippel's Oil, made from blood, bones and other bodily fluids, and to have offered its recipe in exchange for title to his birthplace. Various other Gothic legends are attributed to him (for example, an unhealthy obsession with dissecting animals), but most post-date Mary Shelley's novel and it is unclear whether she would ever have heard of him, even if such tales had been current.

DISGUSTING DREAMS

A mad German scientist whom Mary is more likely to have been aware of was Johann Wilhelm Ritter (1776–1810). A leading figure in the German *Naturphilosophie* movement (*see* page 66), Ritter was the golden boy of German Romantic science until something went horribly wrong. As a brilliant young physiologist at the University of Jena, Ritter's career attracted attention from as far afield as London, where Sir Joseph Banks, President of the Royal Society, kept abreast of his progress. An early pioneer of voltaic pile research, Ritter was one of the first to decompose water electrolytically, pioneered electroplating, and invented a dry-cell battery and an electrical storage battery, all before the age of thirty. His chief claim to fame today rests on his discovery of ultraviolet light.

Ritter's quasi-mystical *Naturphilosophie* led him to believe that the governing principle of the universe was opposition between polarities. Accordingly, when he read of how William Herschel had discovered infrared light by placing a detecting device just beyond the red end of a spectrum of sunlight, he surmised that there must be an opposing type of light at the other end of the spectrum. Herschel had used a thermometer to detect the influence of infrared, but this instrument did not serve at the violet end of the spectrum. Ritter used a chemical salt, silver chloride, which he knew reacted to light by turning black, and which blackened fastest in blue light. Placing vials of silver chloride in the blue part of the spectrum and just beyond it, he discovered that the darkening reaction proceeded more quickly in the latter zone. Ritter dubbed the newly discovered invisible energy "Chemical Rays".

But Ritter's ambition, like that of Frankenstein, went beyond "realities of little worth". According to the poet Novalis (aka Frederick von Hardenburg, a mining engineer): "Ritter is indeed searching for the real Soul of the World in Nature! He wants to decipher her visible and tangible language, and explain the emergence of the Higher Spiritual Forces." Note the parallels with Victor's admiration for "the masters of the science [who] sought immortality and power", and his almost delirious ambition to achieve "more, far more ... treading in the steps already marked, I will pioneer a new way, explore unknown powers, and unfold to the world the deepest mysteries of creation."

Ritter's search for the "Soul of the World" appears to have focused on galvanic responses, and the Irish chemist Richard Chenevix, a Fellow of the Royal Society on a scientific tour of Germany, excitedly wrote to Sir Joseph Banks to inform him of Ritter's progress at Jena. He said that it was Ritter who was doing the "most interesting" work, obtaining "capital results" by using a large voltaic battery that had "a very powerful effect on the animal economy" and yet preserved "the most delicate organs".

In 1804, Ritter moved to Munich to take up a post with the Bavarian Academy of Sciences, and in August of that year Chenevix wrote to Banks to warn that while "Ritter the galvanist is the only man of real talent I have met with", he had fallen under the influence of *Naturphilosophie* and "his head and morals are overturned". In November, Chenevix hinted that Ritter was up to something disturbing and transgressive:

> *I saw him repeat his experiments; and they appear most convincing. Whether there was any trick in them I cannot pretend to say ... It is impossible to conceive anything so disgusting and humiliating for the human understanding as their dreams.*

The nature of these dreams Chenevix does not specify, but Ritter's own posthumous autobiography, *Fragments of a Young Physicist*, suggests that his research followed increasingly occult lines, with water divining and "metal witching" giving way to attempted galvanic reanimation of first animals and then human corpses. There is no evidence to prove these claims, but Ritter's career and mental health collapsed. He neglected his family, locked himself away in his laboratory and eventually went mad, dying penniless and insane in 1810 at the age of just thirty-three. As Richard Holmes, biographer of Shelley and historian of Romantic science, notes, "In other circumstances his memoirs might have been those of young Victor Frankenstein".

Holmes suggests that Ritter's sad and Gothic tale might have been transmitted to Mary and Percy Shelley via any of a number of conduits. It was known to Davy and probably to Lawrence, who had studied in Germany, and it may have been familiar to Polidori. In his introduction to the first edition of *Frankenstein*, Percy specifically mentions "the physiological writers of Germany"; it is highly possible that Ritter was among them.

FATHER FIGURES

The bulk of speculation on real-life inspirations for Frankenstein has centred on figures of whom Mary and Percy would definitely have known. One such candidate is William Lawrence, the surgeon who was such an important influence on Percy and Mary and their thinking on biology, vitalism and materialism. Lawrence was a renowned anatomist and – initially – a fearless advocate of the new mechanistic, materialist strain in scientific thinking. He advocated radical and daring scientific principles that set him at odds with the establishment and threatened to tarnish him as a dangerous and possibly godless materialist – accusations that were levelled at Mary's creation.

Another was Humphry Davy, the brilliant and charismatic Romantic scientist whose enthusiastic demonstrations of electrical science had captured the imagination of Mary's generation (*see* page 49). According to Martin Willis, editor of the *Journal of Literature and Science*, "Davy's association with the vitalist movement (who believed the source of vital power lay within electricity), and his work on apparent harmonies between natural forces, made him revered in romantic circles." As Willis points out, however, Davy features in the novel primarily as Professor Waldman, the most positively portrayed scientist in the book, and indeed Shelley lifted almost wholesale passages from Davy's lectures for one of Waldman's speeches.

Pre-dating Lawrence and Davy as major scientific influences, at least on Percy, was the physician, astronomer and geologist James Lind (1736–1812). When Percy was an unhappy schoolboy at Eton, Lind lived near by and became a friend and mentor. Mary would later record that Percy "has often said ... 'I owe that man far—oh! Far more than I owe my father.'" Lind had travelled the world as a ship's surgeon, and accompanied Sir Joseph Banks of the Royal Society on a scientific expedition to the frozen north in 1772. He was also an enthusiastic investigator of the new electrical science, and his son Alexander recalled the striking scene presented by his father's study: "There were telescopes, Galvanic Batteries, Daggers, Electrical Machines, and all the divers apparatus which a philosopher is supposed to possess."

For many years, including those when Percy was under his wing, Lind collaborated with the London-based Italian physicist Tiberio Cavallo on experiments in galvanism. In 1792 Cavallo wrote to Lind, inquiring "Have you made any dead frogs jump like living ones?", followed a few weeks later by another letter in which he says, "I am glad to hear of your success in the new experiments on muscular motion, and earnestly entreat you to prosecute them to the *ne plus ultra* of possible means". Presumably Lind did not "prosecute these experiments" to quite the limit achieved by Frankenstein, but it is easy to speculate that seeds planted in young Percy's mind would bear fruit in inspiration the poet provided to his lover as she conceived the *ne plus ultra* of galvanic ambition.

CLOSE AT HAND

Literary analysis of *Frankenstein* tends to downplay as overly reductive and simplistic the notion that any single real-life figure was the model for Victor. Instead more weight is given to the influence of the most influential men in Mary Shelley's life: her father and her lover/husband. The novel was dedicated to William Godwin, and one reading of the novel paints him as Victor and Mary as the monster; like him, she has to grow up without a mother, only to be cruelly rejected by her arrogant father. Meanwhile literary critic Professor Mary Lowe-Evans says that "No discussion of *Frankenstein*'s men would be complete without fully acknowledging the role Percy Shelley played in their formulation ... there seems little doubt that he inspired certain aspects of Victor ..." Lowe-Evans lists the relevant aspects of his "volatile and unusually complex persona", including his "interest in science, his tendency to become obsessively enthusiastic about whatever project he had currently under way, his need for periodic isolation, his powers of persuasion, his alternating fits of anguish and joy ..." She points out that Percy had even used "Victor" as a pen-name.

A PHILOSOPHER OF THE RAREST STAMP

Galvanic experiments of almost alchemical nature were the hallmark of a lesser-known candidate for the "real Frankenstein": Andrew Crosse. He is chiefly known for the affair of the "electric mites" ("Crosse's Acari"), which post-dates the composition of *Frankenstein* (*see* page 88). However, his reputation as a "mad scientist" extended beyond this notoriety. A profile in *Chamber's Journal* described him as "a philosopher of the rarest stamp, one disposed to pursue nature into her coyest recesses, and wring from her her most mystic secrets". His research alarmed local residents, who christened him "the thunder and lightning man" and "the Wizard of the Quantocks" (in reference to the nearby hills), as he strung copper wires around his estate and experimented on atmospheric electricity. An account of a visit to his laboratory describes a scene fitted to James Whale's 1931 film version of the novel:

you are startled in the midst of your observations by the smart crackling sound that attends the passage of the electrical spark; you hear also the rumbling of distant thunder. The rain is already splashing in great drops against the glass, and the sound of the passing sparks continues to startle your ear. Your host is in high glee, for a battery of electricity is about to come within his reach a thousand-fold more powerful than all those the room strung together. You follow his hasty steps to the organ-gallery, and curiously approach the spot whence the noise that has attracted your notice. You see at the window a huge brass conductor, with a discharging rod near it passing into the floor, and from the one knob to the other, sparks are leaping with increasing rapidity and noise, rap, rap, rap – bang, bang, bang...Nevertheless, your host does not fear. He approaches as boldly as if the flowing stream of fire were a harmless spark.

Could Mary have been aware of this figure? Crosse gave a lecture on his electrical experiments in 1814, and there is conjecture that Percy and Mary attended it, so it is possible that he helped to inspire the portrayal of Victor.

MAD SCIENCE: FRANKENSTEIN AS ARCHETYPE AND INSPIRATION

Though many critics contend that the novel itself is of debatable literary virtue, none will deny that Mary Shelley created something much grander and more powerful than a simple, or even a complex and sophisticated, fiction. *Frankenstein* became a modern myth, one of the few to have been created in the Western canon since classical times. Its tropes and clichés, interpretations and misinterpretations have informed discourse around science and scientists ever since.

The moral of the novel, if there is one, is heavily contested. Is Frankenstein punished because his presumption trespasses on divine right, because any technology is ultimately destructive, or because he fails in his duty to the being he has created, through a combination of cowardice, weakness and arrogance? The power of a myth is that it is open to multiple, possibly infinite interpretations and uses, and so the myth of Frankenstein has found application to an extraordinary range of scientific advances. There are, however, three areas in which the analogy has been most forcefully drawn, to greatest effect: the atomic bomb, genetic engineering and intelligent machines.

DESTROYER OF WORLDS

The Manhattan Project to create the first atomic bomb was initiated in the full light of knowledge about what might be unleashed. As early as 1913, inspired by reports about research into radioactivity, H. G. Wells had predicted atomic bombs in his novel *The World Set Free*, and the book in turn helped to inspire a new generation of atomic scientists to consider the possibility. One particularly far-sighted physicist, the Hungarian Leo Szilard, worked tirelessly to encourage those nations aligned against the Axis Powers to develop a bomb before they did. Under the direction of Italian physicist Enrico Fermi, a demonstration nuclear reaction pile was constructed, in which neutrons would be released through the splitting – or fission – of uranium atoms. These neutrons would trigger further fission, setting off a chain reaction with potential to release a vast flood of energy.

In 1942 the pile was activated and a successful chain reaction initiated, and quickly stifled. Szilard recalled, "I shook hands with Fermi and I said that I thought this day would go down as a black day in the history of mankind. I was quite aware of the dangers ..."

The dangers would become still more apparent with the first detonation of an atom bomb: the Trinity test of July 16, 1945. This was the culmination of the colossal Manhattan Project, overseen by physicist Robert Oppenheimer. Biologist Leonard Isaacs, in his essay "Creation and Responsibility in Science: Some Lessons from the Modern Prometheus", draws explicit parallels between Oppenheimer and Frankenstein, comparing their mystical streaks and scientific educations. On coming face to face with his creation Oppenheimer experienced profound ambivalence, famously quoting a line from Hindu scripture: "I am become death, destroyer of worlds". Oppenheimer's sense that the creation of the bomb was a monstrous, almost blasphemous act, deepened with time: "In some sort of crude sense which no vulgarity, no humour, no overstatements can quite extinguish," he said in 1947, "the physicists have known sin; and this is a knowledge which they cannot lose".

Isaacs further develops the parallel, suggesting that it was the nuclear arms establishment as much as the bomb itself that was the monster, and likens the monster's demand for Victor to build him a bride to the insistence of the military establishment that atomic physicists should develop the more powerful H-bomb (a fusion bomb). The link between the bomb and the novel was explicitly made by James B. Conant, following a meeting of the Atomic Energy Commission with Oppenheimer's advisory committee on the hydrogen bomb, when he opposed the effort on the basis that "*we built one Frankenstein ...*" Oppenheimer concurred, and as a result his career was ruined. This monster too had destroyed its creator, but unlike Mary Shelley's version, it refused to disappear and destroy itself.

LITTLE BIOLOGICAL MONSTERS

More direct analogy has been offered by the science of genetics, which, like Victor, seeks to master the secrets of life and potentially generate new beings. From the start this project was conceived in Frankensteinian terms. James Watson, the American microbiologist who worked with English biophysicist Francis Crick to discover the structure of DNA, said that he "daydreamed about discovering the secret of the gene", and that "a potential key to the secret of life was impossible to push out of my mind". Watson recalled that when they had completed their model of the double helix on 28 February 1953, Crick went to the pub and announced that they had "found the secret of life".

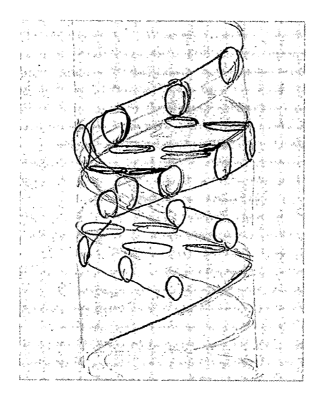

Francis Crick's original sketch of the structure of DNA made in 1953; Crick told friends at the pub that he had "found the secret of life".

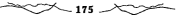

The structure revealed by Crick and Watson clearly indicated that genetic information is encoded in the sequence of molecular components known as bases. It was not long until biologists had worked out precisely how the sequence of bases encoded the production of proteins; scientists now believed they could "read the book of life", and technologies collectively known as recombinant DNA meant that the book might be edited or even rewritten. Molecular biologist Robert Sinsheimer observed, "For the first time in all time, a living creature understands its origin and can undertake to design its own future". By the 1970s the emergence of this new science – genetic engineering – set off a firestorm of ethical debate. Sinsheimer wondered, "Do we want to assume the basic responsibility for life on this planet – to develop new living forms for our own purpose? Shall we take into our own hands our own future evolution?" (*See* page 187 for more on synthetic life.)

Biochemist Erwin Chargaff, whose work had helped Crick and Watson reveal the double helix, was among the first to raise the spectre of Frankenstein in a 1976 letter to the journal *Science*:

> *Is there anything more far-reaching than the creation of new forms of life? ... Have we the right to counteract, irreversibly, the evolutionary wisdom of millions of years, in order to satisfy the ambition and the curiosity of a few scientists?*

Chargaff was particularly concerned at the use of the bacteria *E. coli* as the vessel for most of the genetic engineering research:

> *If Dr. Frankenstein must go on producing his little biological monsters – and I deny the urgency and even the compulsion – why pick* E. coli *as the womb? ... The hybridization of Prometheus with [ancient Greek arsonist] Herostratus is bound to give evil results.*

ATTACK OF THE FRANKENFOODS

Since then, *Frankenstein* has become synonymous with anxiety over developments in genetics, and biology in general. This had been foreseen in 1924 by British biologist J. B. S. Haldane:

> There is no great invention, from fire to flying, which has not been hailed as an insult to some god. But if every physical and chemical invention is a blasphemy, every biological invention is a perversion.

In the 1990s, when the products of genetic engineering finally began to emerge from the laboratory in the form of genetically modified (GM) foods, the "Franken-" hype went into overdrive. It began with a letter to the *New York Times* on 2 June 1992, in which English professor Paul Lewis protested against the advent of GM food: "If they want to sell us Frankenfood, perhaps it's time to gather the villagers, light some torches and head to the castle". Since then, the media has warned of an extraordinary range of monstrous organisms, including "frankengrass", "frankenbugs", "frankenseed", "frankenveggies", "frankencorn", "frankenchicken" and "frankenfish", all raised on "frankenfarms". This animus is motivated by what American bioethicist Leon Kass called "the wisdom of repugnance" – the very same motive that led Frankenstein and everyone else who encountered him to condemn and recoil from the monster.

The bold predictions of 1970s genetic engineers have remained largely unfulfilled. However, the development of the CRISPR gene-editing technique, which promises to make genetic engineering easier and better, has triggered a fresh wave of anxieties in which Frankenstein is invoked. To give just one example, a 3 April 2017 article about CRISPR in the *Indian Telegraph* newspaper used the headline "Editing Frankenstein".

MAN VERSUS MACHINE

*F*rankenstein was written just as the Industrial Revolution was accelerating. The previous century had produced major innovations in technology, finance and labour, leading to the emergence of factories powered by steam engines driven by coal. The nineteenth century would see these innovations feed through into colossal and often traumatic shifts in demography, lifestyles and inequality. Anxiety and anger around the pace and impact of these changes often focused on the most visible symbols: the machines. Just a few years before Mary Shelley wrote *Frankenstein*, England had been convulsed by a wave of violent machine-breaking protests known as the Luddite riots, in which artisan weavers, angered at the threats to their livelihoods posed by the new factory systems and the growing imbalance between labour and capital, broke into factories to destroy weaving machines and power looms. Luddite riots began in Nottinghamshire in 1811 and spread throughout the country, particularly in the north.

Inevitably, there have been socio-economic readings of the novel as a parable about arrogant capitalists creating a monstrous system that runs out of control, or simply in the more general sense of warning of the dangers of technology unleashed on an unprepared world. Since the emergence of science fiction and the increasing cultural penetration of the idea of robots and machine intelligence, this analogy between Frankenstein's monster and the machine threat has been repurposed and powered up.

The modern-day ubiquity of the close identification between Frankenstein's monster and robots is worth questioning, because much of the point of Victor's creation is precisely that he is not a machine, but an all too organic thing of flesh and blood. The key theme that connects them is that of the automaton, which, as we have seen (*see* Chapter 4), haunted the Romantic imagination in the era of the mesmeric trance, somnambulism, cunning automata and the mechanistic threat to the traditional transcendental dispensation.

This connection finds its greatest support not in the novel – in which the monster is autonomous rather than automatic – but in the stage and screen adaptations most responsible for propagating the Frankenstein myth. Once the monster has been transformed into a lumbering mute his humanity is obscured, leaving an automaton. The seminal 1931 Universal Pictures adaptation cemented this version of the monster, with Boris Karloff's flat head, metal neck bolts (actually intended as points to which electrodes could be attached) and mechanical gait. To assist Karloff in producing his stilted, mechanistic movements, he wore a metal rod next to his spine, and it is thought that the designers on the film may have drawn inspiration for their version of the monster from "Televox", a mechanical man developed by engineers at Westinghouse in the 1920s.

Westinghouse's rudimentary robot Televox, which could respond to commands given remotely and even talk.

A RACE OF DEVILS

Accordingly, the most prolific and visible descendants of the monster have been androids – humanoid robots – in the countless stories and scenarios in which machine beings become autonomous and run amok, or at least threaten their creators. However, the analogy with Frankenstein's monster has in some ways come back to its roots in its most contemporary manifestation.

In Chapter 20, Victor is tortured with doubt about the female companion he has agreed to make for the monster, foreseeing dreadful consequences since "one of the first results of those sympathies for which the daemon thirsted would be children, and a race of devils would be propagated upon the earth who might make the very existence of the species of man a condition precarious and full of terror". This grim prediction is startlingly prescient of modern fears about machines, which take two forms: self-replicating machines and the "grey goo" threat; and the Singularity.

The "grey goo" scenario was first proposed by Dr Eric Drexler, one of the fathers of nanotechnology, in his 1986 book *Engines of Creation*. Drexler described a hypothetical scenario in which a nanomachine – a tiny machine with parts as small as molecules – is designed to be self-replicating: able to assemble another copy of itself from building blocks it is able to scavenge. Such a machine would reproduce exponentially: one machine would copy itself to produce two, each of these would copy itself to produce four, four would become eight, 16, and so on. It would take only 20 generations to produce a million, and only 30 to produce a billion. The only limit on reproduction would be resources; if the machines ran out of building blocks they would not be able to reproduce further.

This is essentially how a virus works, making use of the resources of a host to replicate exponentially until the resources are exhausted. Viruses, however, have to be highly specific in order to get into their hosts, and also tend to be fairly fragile, with difficulty surviving for long outside of very specific conditions (that is, host cells). This limits their range and spread and prevents them from taking over the planet. Nanomachines might not be subject to the same constraints. They might be specifically designed to be robust and generic, allowing them to operate in a wide range of conditions, while their artificial nature might mean that immune systems would not be able to cope with them.

If the nanomachine were sophisticated enough, it could derive the building blocks it needed from more complex substances by breaking them down. At this point it would become incredibly dangerous, because it would effectively be able to "eat" other substances. Since self-replicating nanomachines would most probably be constructed of organic molecules, the "targets" for these machines would be all organic matter, which includes all life on Earth. Assuming the machines were efficient and robust, they would probably replicate fast enough to overwhelm any defences that could be mustered, and would eventually render all organic matter – including people, animals, plants, plankton, micro-organisms, soil, plastics, wood, fabrics and even fossil fuel reserves – into nanomachines. All that would be left would be a mass of nanomachines in water, which would resemble a grey goo covering the face of the planet.

SUMMONING THE DEMON

The Singularity is the hypothetical climax of an altogether different type of machine evolution. While nanobots are mindless physical entities, artificial intelligence – or AI – involves minds without bodies. Ever since the advent of electronic computers at the end of the Second World War, computer pioneers have been speculating about AI, and in 1965 British mathematician Irving John Good published his seminal paper "Speculations Concerning the First Ultraintelligent Machine", anticipating the eventual existence of superhuman intelligence:

> Let an ultraintelligent machine be defined as a machine that can far surpass all the intellectual activities of any man however clever. Since the design of machines is one of these intellectual activities, an ultraintelligent machine could design even better machines; there would then unquestionably be an "intelligence explosion," and the intelligence of man would be left far behind. Thus the first ultraintelligent machine is the last invention that man need ever make.

What Good called the "intelligence explosion" was christened the Singularity by computer scientist and sci-fi author Vernor Vinge, who derived the term from a description by atomic bomb scientist Stanislaw Ulam of a conversation he had had with Hungarian-American mathematician, physicist and computer scientist John von Neumann:

> *One conversation centred on the ever accelerating progress of technology and changes in the mode of human life, which gives the appearance of approaching some essential singularity in the history of the race beyond which human affairs, as we know them, could not continue.*

Stephen Hawking in his office in 1982; Hawking is one of the influential voices warning of the dangers of "Frankenstein AI".

While some view the Singularity as a millenarian event leading to a utopian future of eternal life, others warn darkly that true AI poses an existential threat and that the Singularity could mark the end of the human race. The Jeremiahs – or Cassandras, depending on assessment of their accuracy – include prominent figures in science and technology such as Stephen Hawking and Elon Musk. Hawking, for instance, co-authored an article that warned, "Success in creating AI would be the biggest event in human history. Unfortunately, it might also be the last, unless we learn how to avoid the risks ... Whereas the short-term impact of AI depends on who controls it, the long-term impact depends on whether it can be controlled at all."

Musk has labelled AI "our biggest existential threat", calling it "potentially more dangerous than nukes", and warning "we do not have long to act. Once this Pandora's box is opened, it will be hard to close." In even more emotive language, echoing the Faustian roots of the Frankenstein myth, Musk has liked AI to "summoning the demon ... In all those stories where there's the guy with the pentagram and the holy water, it's like – yeah, he's sure he can control the demon. Doesn't work out."

HUMAN NATURE

One problem with these prophecies is that many view AI as a mirage, one of those much-heralded technologies that are "thirty years away and always will be". But evidence from rudimentary machine intelligences that have already been released into the world is not promising. One of the most infamous examples was the release by Microsoft of a chatbot named Tay. ai, an AI system designed "to entertain and engage" 18-24-year-olds. The idea was that the chatbot would learn to engage people through "casual and playful conversation", using what Microsoft called "conversational understanding" of tweets. In less than 24 hours users managed to make the chatbot retweet and even invent a variety of racist, misogynistic and anti-Semitic remarks, and it had to be taken offline.

If there is a lesson to be drawn from the Tay debacle, it is that the human element may pose the greatest danger; certainly an obvious reading of *Frankenstein*. A related argument is marshalled by Yann LeCun, Director of AI Research at Facebook and Professor at

NYU, who downplays the threat from AI on the basis that machine intelligence would be free of the sins and vices of humans:

> A lot of the bad things humans do to each other are very specific to human nature. Behavior like becoming violent when we feel threatened, being jealous, wanting exclusive access to resources, preferring our next of kin to strangers, etc., were built into us by evolution for the survival of the species. Intelligent machines will not have these basic behaviors unless we explicitly build [them] into them.

In effect LeCun is arguing that the Frankenstein analogy for AI is a category error – the monster is dangerous precisely because of his humanity; creations that do not partake of this humanity will be less dangerous, not more.

REANIMATION AND FUTURE SCIENCE

Mary Shelley was both artfully and necessarily vague about the science that made her monster possible. Relating his epochal discovery – possibly the greatest ever made – of the scientific key to unlock death and give life to inanimate tissue, Victor retreats into rather woolly language, speaking of "a sudden light" and an "astonishing secret". There are no more clues to be found in Victor's creature-building laboratories, where Shelley mentions only "the instruments of life" or "chemical apparatus".

NO CURE FOR DEATH

The first hundred years following the publication of the novel made its prospect less plausible, since the same advances in biology that dissipated the notion of life energy or a vital spark also elucidated the processes of organic decay that attend on death. It is now known, for instance, that almost as soon as a cell stops receiving sufficient oxygen and nutrients to function, its internal machinery triggers a process known as necrosis, or cell death. This is where damage or insult to the cell causes it to lose control of or release corrosive enzymes and other chemicals, which go on to generate more of themselves, attack the integrity of the cell, and release these corrosive chemicals into the tissues of the body. The incredible complexity of the cell, which constitutes the phenomenon of life, also means that damage such as this almost immediately becomes irreparable.

Mary Shelley's contemporaries, including the most well informed physicians and natural philosophers, might have assumed that it was, at least in principle, plausible for dead and even rotting tissue to be re-animated. Indeed some theories of spontaneous generation insisted on decay and corruption (see page 77). But by the mid-twentieth century it was clear that Victor's choice of "raw material", sourced from "charnel houses ... the dissecting room and the slaughterhouse", effectively torpedoed any claims to scientific plausibility that might be made for the novel.

So the knowledge necessary to give a plausible scientific account of the monster's creation did not exist in Mary's day. What about today? Could technologies being pursued today finally realize Frankenstein's dream of triumph over death? In fact there are a suite of near and far future technologies intended to accomplish just this.

SPARE PARTS

If repairing dead tissue is far-fetched due to the rapidity and irreversibility of cell death, perhaps tissue can be replaced? There is a wide range of mechanical and organic replacement body parts and tissues already available, and many more in development. The earliest prosthetics date back to ancient times – an Egyptian mummy has been found with a prosthetic big toe, for instance, while the ancient Roman general Marcus Sergius was said to have been fitted with an iron replacement for his right hand. The metal prosthetic was functional enough to allow him to return to battle against the Carthaginians in the Second Punic War (218–201 BCE), and he enjoyed a long career.

Cutting-edge modern prosthetics are bionic, and there are now those able to interface with nerves, either in stumps or in the brain, to allow direct control by the wearer. Other bionic body parts already available include cochlear implants, which artificially simulate hearing function, and retinal implants that can send simple visual stimuli through the optic nerve. Moving closer to the Frankenstein scenario, researchers are developing organs and body parts grown to order. Combining stem cells and 3-D printing, researchers have shown it is possible to build a scaffold or matrix of organic or inorganic material, around which stem cells then grow functional tissue. This approach has already yielded human skin, ears, muscle tissue including heart tissue, and bones. Animal hosts could potentially be used to grow whole organs, and there is also work under way to re-programme the cells in animal organs for transplant, so that human recipients will not reject them. A related possibility, with terrible ethical implications that have spawned a mini-genre of science fiction, is cloning humans in order to farm them for organs. Dolly the sheep, born in 1996, proved that it is possible to clone an adult mammal: she was created using a nucleus extracted from a cell from an adult sheep's mammary gland, which was then added to a donor egg from which the nucleus was removed. Dolly was explicitly compared to Frankenstein's monster.

The 1997 "Vacanti mouse" had an ear-shaped mould of bovine cartilage cells implanted under its skin, around which grew natural cartilage. Similar methods using stem cells might produce replacement human organs and body parts.

MADE TO ORDER

As discussed on page 177, the emergence of genetic engineering was attended by frequent use of Frankenstein metaphors and rhetoric. But analogies between genetically engineered organisms and the monster are often very loose: the vast majority of genetic engineering involves editing existing genomes – as if, say, Victor had swapped his friend Clerval's arms for better ones. The closest that biology has actually come to creating an entirely artificial living being is probably Craig Venter's synthetic life form, JCVI-syn, also known as Synthia. Venter is a controversial American biotech scientist who led a team that sequenced a human genome (his own) in the 1990s and has since focused mainly on the field of synthetic life. In 1995 Venter's team sequenced the genome of *Mycoplasma genitalium*, a sexually transmitted microbe with the smallest genome of any known free-living organism; it has just 470 genes. This led him to speculate on the minimal genome necessary to support life, and he and his collaborators Clyde Hutchison and Hamilton Smith developed new techniques to synthesize an entire genome of their own design.

Initially, they simply made a synthetic copy of the *M. genitalium* genome, switching to the related but much faster-growing *Mycoplasma mycoides* bacterium and doing the same. In 2010 Venter and his team synthesized an entire chromosome for *M. mycoides*, adding genetic coding spelling out their names and some famous quotes, among them a version of Richard Feynman's assertion, "What I cannot create, I do not understand." Thus genetically "watermarked", the synthetic chromosome was added to a bacterium that had its own genome removed, and the resulting cell proved viable. Venter christened it JCVI-syn 1.0, and announced that he had created life, ensuring that he is the modern scientist most compared to Frankenstein.

The press nicknamed the synthetic organism Synthia, but Synthia was simply a copy of an existing organism. Still seeking to synthesize a true minimal genome, Venter and his team experimented with knocking out the genes from the *M. mycoides* genome until they had identified all those that could be left out but still give a viable, self-replicating organism. In 2016 they synthesized a 473-gene chromosome and implanted it into a genome-deleted bacterial cell to create JCVI-syn 3.0, an organism with the absolute minimum number of genes required to support life. This microbe has to have nearly all its nutrients supplied to it, since there are almost none it can digest or make. One of its most extraordinary features is that 149 of its genes encode unknown functions, highlighting just how far there is left to go for scientific understanding of the most basic biology of life. Nonetheless Synthia 3.0 is probably the closest thing to a real-life Frankenstein's monster: a living organism with a unique and patented genome compiled by humans, although even in this case the "body" or host for this genome was supplied by nature, as scientists are still not able to construct a working cell from scratch.

GETTING A HEAD

If Victor Frankenstein had used fresh body parts from living donors, there is a chance that his experiment might really have worked. This is, in effect, the contention of controversial Italian neurosurgeon Sergio Canavero, who made headlines in 2016 by promising to perform the first human head transplant. An impressive list of organs and body parts have been successfully transplanted: hearts, livers, kidneys, lungs, uteruses, voice boxes, tongues, penises, hands and even faces. Hand and face transplants involve incredibly sophisticated microsurgery to join together diverse and delicate tissues including muscle, skin, bone, tendons, cartilage, blood vessels and nerves, although there are important differences between peripheral nerves and spinal ones, which cannot currently be made to regrow.

This is why historical trials of head transplantation in animals did not involve attempting to reattach spinal cords. Soviet experiments on dogs by Vladimir Demikhov in the 1950s showed that the head and forelimb of one dog could be surgically grafted onto the body of another and survive, at least for a few days. In 1970, Robert White led a team that attached the severed head of one rhesus monkey to the body of another. The head regained consciousness and was able to blink and look around, but the monkey was paralysed and died after eight days due to tissue rejection.

Canavero insists that he can reattach severed spinal cords. He claims that application to severed spinal nerves of an organic glue, polyethylene glycol (PEG), can induce nerve cells to fuse together, and he has specifically cited *Frankenstein* as inspiring a breakthrough in his approach. He says that reading the novel encouraged him to think of using electrical stimulation in concert with PEG to achieve successful nerve fusion: "Electricity has the power to speed up regrowth... Bing bang bong you have the solution ..."

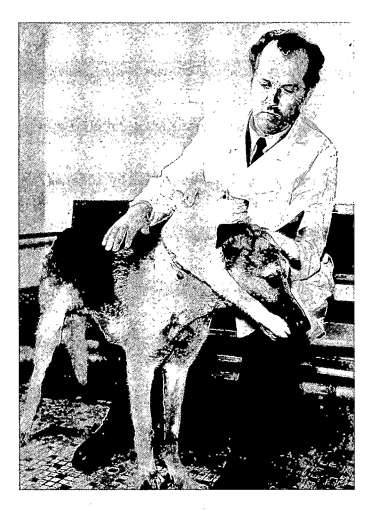

Controversial experiments by Soviet scientist Vladimir Demikhov showed that a puppy's upper body could be attached to another dog's circulatory system, as in this 1959 trial.

Canavero's Chinese collaborator Xiaoping Ren claims to have achieved successful demonstrations in mice and on a monkey, and Canavero made headlines around the world with an announcement that in 2017 he would transplant the head of a wheelchair-bound Russian man, Valery Spiridonov, onto the body of a brain-dead donor. A flood of opprobrium followed his brash pronouncements, with critics branding Canavero "insane", "delusional" and "corrupt" and describing him as having "a knife and a mad glitter in his eye". His plan was called a "medical horrorshow" of "Bond-villain insanity", and the name Frankenstein was inevitably flung around, not least by Canavero, who frequently compares himself to the fictional monster-maker and uses similar language. He has, for instance, said that he wants to "breach the wall between life and death". Ren, too, has been tarred with this brush, labelled "China's Frankenstein" by a Hong Kong newspaper.

Almost every expert thinks that a human head transplant has no chance of success, that spinal cord fusion using PEG is not properly demonstrated, and that any patient involved will almost certainly die – or at best, live for a short time and be paralysed from the neck down. Also it should be noted that there are suspicions that Canavero has been staging a hoax and that the whole thing is part of a marketing stunt for a computer game named *Metal Gear Solid 5: The Phantom Pain*, which features a character apparently modelled on Canavero. At the time of writing it is not clear if Spiridonov (also implicated in the marketing stunt theory) is still involved, or if Canavero's claims of successful tests on corpses will lead to an attempt with living donors.

Canavero aside, should the principle of head transplantation be considered unthinkable? The idea triggers a strong emotional or gut reaction, an instinctive disgust akin to the horror provoked by the creature. But the "wisdom of repugnance" was also once applied to other organ transplants. The first human heart transplant, performed in 1967, was surrounded by Frankenstein rhetoric, and the recipient, Louis Washkansky, reportedly said on waking from the surgery, "I am the new Frankenstein" (presumably he meant the monster). Richard Lawler, who performed the first kidney transplant, may not have faced an angry torch-wielding mob, but he did face social and professional censure. Hand and face transplants also faced a lot of opposition when originally mooted. Perhaps the concept of head transplants should not be rejected out of hand.

HEAD OR BODY TRANSPLANT?

Arguably the procedure Canavero discusses above should be called a body transplant, because the "person" that is benefiting from/receiving the transplant is the one supplying the head – at least working on the assumption that a person is their brain (though see the box "Body of Knowledge" on page 192). On the other hand, the head contributes less than 10 per cent of the mass of the body, and it would be the body that attempts to reject the head after the surgery.

Valery Spiridonov, once a candidate for a head (body?) transplant, although doubts have been raised about the plausibility and legitimacy of the whole affair.

BODY OF KNOWLEDGE

The prospect of head transplants raises a host of fascinating questions that could equally well apply to Frankenstein's monster: to what extent did his personality, emotions and moods, behaviour and so on rely on the brain Victor had given him, as opposed to the body? Hormones and other chemical signals from the body profoundly affect brain function and cognition; for example, there is evidence that gut bacteria secrete chemicals that change mood and behaviour. Many procedural memories, such as playing an instrument or riding a bicycle, involve "muscle memory", and it is not clear to what extent muscle memories are encoded by or depend upon the actual muscles. If so much of psychology depends on being embodied, what would be the impact of having a chimeric body composed of mismatched parts from multiple sources? Whose muscle memories did the creature have? Did his aggression and capacity for violence derive from his brain or from his endocrine organs, such as his testes or adrenal glands?

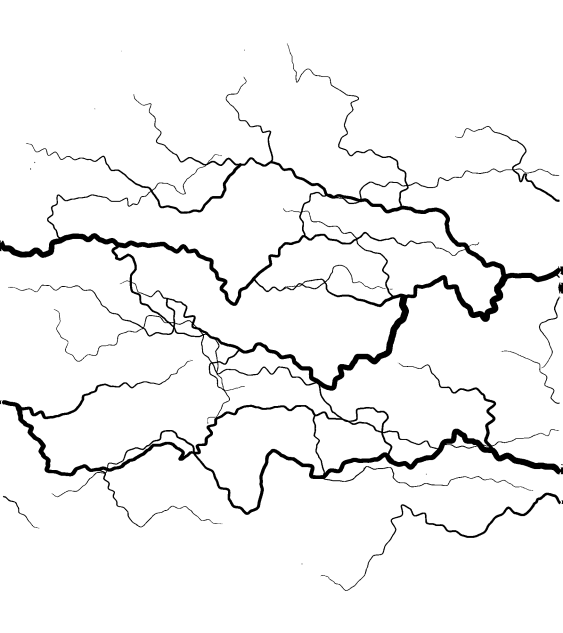

FURTHER READING

Alan Rauch; "The Monstrous Body of Knowledge in Mary Shelley's Frankenstein"; *Studies in Romanticism*, 14 (1995), 227-53

Alison Winter; *Mesmerized: Powers of Mind in Victorian Britain*; University of Chicago Press, 1998

Anne Mellor; *Mary Shelley: Her Life, Her Fiction, Her Monsters*; Methuen, 1988

Brian Scott Baigrie; *Electricity and Magnetism: A Historical Perspective*; Greenwood Press, 2007

Chet Van Duzer; "The Mythic Geography of the Northern Polar Regions: Inventio fortunata and Buddhist Cosmology". Culturas Populares. Revista Electrónica 2; May/August 2006; www.culturaspopulares.org/textos2/articulos/duzer.htm

Claire Tomalin; *The Life and Death of Mary Wollstonecraft*; Penguin, 1992

David Gubbins and, Emilio Herrero-Bervera [Eds]; *Encyclopedia of Geomagnetism and Paleomagnetism*; Springer, 2007

David Welky, *A Wretched and Precarious Situation: In Search of the Last Arctic Frontier*; WW Norton, 2016

Hana Akselrod, Mark W. Kroll, and Michael V. Orlov; "History of Defibrillation"; in I. R. Efimov et al. (Eds), *Cardiac Bioelectric Therapy: Mechanisms and Practical Implications*; Springer Science+Business Media, 2009

Helen MacDonald, "Legal Bodies: Dissecting Murderers at the Royal College of Surgeons, London, 1800–1832", *Traffic: An Interdisciplinary Postgraduate Journal*, No. 2, 2003

James Bieri; Percy Bysshe Shelley: *A Biography: Youth's Unextinguished Fire, 1792–1816*; University of Delaware Press, 2004

Janis McLaren Caldwell; *Literature and Medicine in Nineteenth-Century Britain: From Mary Shelley to George Eliot*; Cambridge University Press, 2004

Joel Levy; *The Infinite Tortoise: The Curious Thought Experiments of History's Great Thinkers*; Michael O'Mara, 2016

Len Fisher; *Weighing the Soul: Scientific Discovery from the Brilliant to the Bizarre*; NY: Arcade Publishing, 2011

Leonard Isaacs; "Creation and Responsibility in Science: Some Lessons from the Modern Prometheus"; in *Creativity and the Imagination: Case Studies from the Classical Age to the Twentieth Century*; University of Delaware Press, 1987

Linda Simon; *Dark Light: Electricity and Anxiety from the Telegraph to the X-Ray*; Houghton Mifflin Harcourt, 2005

Martin Willis; "Frankenstein and the Soul"; *Essays in Criticism*, 45:1 (Jan. 1995), 24-35

Mary Lowe-Evans; "The Groomsmen", *Bloom's Major Literary Characters: Frankenstein*; Chelsea House Pub, 2004

Michael Worboys; *Spreading Germs: Disease Theories and Medical Practice in Britain, 1865–1900*; Cambridge University Press, 2000

Per Schelde; *Androids, Humanoids, and Other Science Fiction Monsters: Science and Soul in Science Fiction Films*; NYU Press, 1994

Philip Ball; *Unnatural: The Heretical Idea of Making People*; Vintage, 2012

Richard Holmes; *Age of Wonder*; Harper Press, 2008

Seamus Deane, *The French Revolution and Enlightenment in England 1789-1832*; Harvard University Press, 1988

Tim Marshall; *Murdering to Dissect: Grave-robbing, Frankenstein and the Anatomy Literature*; Manchester University Press, 1995

INDEX

(page numbers in italic type refer to illustrations and photographs)

CREDITS AND ACKNOWLEDGEMENTS

The publishers would like to thank the following sources for their kind permission to reproduce the pictures in this book.

13. Apic/Getty Images, 18. DeAgostini/Getty Images, 20. Wellcome Library, London, 21. Wikimedia Commons, 23. Fototeca Gilardi/Getty Images, 25. Mary Evans Picture Library, 26. Falkensteinfoto/Alamy Stock Photo, 28. World History Archive/Alamy Stock Photo, 31. Classic Image/Alamy Stock Photo, 35. Universal History Archive/Getty Images, 36. Granger Historical Picture Archive/Alamy Stock Photo, 39. Granger/REX/Shutterstock, 44. Time Life Pictures/Mansell/The LIFE Picture Collection/Getty Images, 54. Public Domain, 56. Wellcome Library, London, 58. Paul Fearn/Alamy Stock Photo, 63 & 68. Wellcome Library, London, 79. Public Domain, 81. Universal History Archive/UIG/Getty Images, 84. Granger/REX/Shutterstock, 85. Smithsonian Institute, 87 & 89. Chronicle/Alamy Stock Photo, 93 & 95. Public Domain, 96 & 98. Wellcome Library, London, 101. CORBIS/Corbis via Getty Images, 106. Design Pics Inc/REX/Shutterstock, 111. Science History Images/Alamy Stock Photo, 113. Interfoto/Alamy Stock Photo, 117. Public Domain, 124. Public Domain, 125. Wellcome Library, London, 129 & 133. Public Domain, 136. Three Lions/Getty Images, 143. Courtesy Picture Collection, The Branch Libraries, The New York Public Library, 144. Granger/REX/Shutterstock, 145. Courtesy of Osher Map Library, 148 & 151. Public Domain, 154. National Library of Norway, 157. Oxford Science Archive/Print Collector/Getty Images, 159. Public Domain, 163. © SZ Photo/ Bridgeman Images, 166. Ann Ronan Pictures/Print Collector/Getty Images, 175. Science History Images/Alamy Stock Photo, 179. Underwood Archives/ Getty Images, 182. Bettmann/Getty Images, 186. Private Collection, 189. Keystone-France/Gamma-Keystone via Getty Images, 191. Sergei Fadeichev\ TASS via Getty Images, 193. Shutterstock.com

Every effort has been made to acknowledge correctly and contact the source and/or copyright holder of each picture and Carlton Books Limited apologises for any unintentional errors or omissions, which will be corrected in future editions of this book.

Thanks to Anna Darke.